THE METAVERSE

ALL YOU NEED TO KNOW ABOUT THE FUTURE OF INVESTING

Amy Satpalson

Pharos Books

Hardcover ISBN: 978-93-95862-79-0
Paperback ISBN: 978-93-95862-78-3
eBook ISBN: 978-93-95862-77-6

©Publisher

Publisher: Pharos Books (P) Ltd.
Plot No.-55, Main Mother Dairy Road
Pandav Nagar, East Delhi-110092
Phone: 011-40395855, +4049916623
WhatsApp: +91 8368220032
E-mail: sales@pharosbooks.in
Website: www.pharosbooks.in
First Edition: 2022

Printed By: Sushma Book Binding House, Okhla
Industrial Area, Phase II, New Delhi-110020

The Metaverse
By Amy Satpalson

CONTENTS

Introduction

After *Facebook* was renamed Meta, there was a fresh buzz around the term *"Metaverse,"* but no one knew what it meant. For many years, people have speculated on how the Internet and technology would progress. When we talk about various evolutionary stages of the Internet, we commonly refer to Web1, Web2, Web3, and so on, and now we're talking about the Metaverse. But forget the definitions for a moment and consider this: Consider the possibility of forming your ideal alter-ego in a virtual environment where you have complete control over every element. You can do anything you want, own whatever you want, and your choices are nearly unlimited.

Such a scenario was frequently depicted in science fiction films or television programs. However, it may become a reality soon since there is a lot of buzz in the business sector about getting every layer, technology, and protocol ready to build "The Metaverse." Most debates about what the Metaverse contains begin to stagnate at this point. We have a hazy picture of what objects exist in what we would term the Metaverse, and we know which corporations are investing in the concept, but we still don't know what it is. Facebook—sorry, Meta, I still don't understand it—believes it will contain fictional residences to which you can invite all of your pals to hang around. Microsoft appears to believe that virtual meeting rooms may be used to teach new recruits or converse with faraway coworkers. This book will teach you more about Metaverse and how it works. But what's the big deal about the Metaverse? And what does it have to do with NFTs, Blockchain, or living on the Internet 24 hours a day, seven days a week? Let's get started since there's a lot to unpack.

1

INTRODUCTION TO NFTS

History of the Metaverse

What is the Metaverse?

Metaverse is a term formed by combining two different words, 'Meta' and 'Universe'. It is essentially a digital space where digital objects and things represent digital people. It is an idea that should consolidate parts of a few innovations, including social media, augmented reality (AR), virtual reality (VR), internet gaming, and cryptocurrencies. And allow people to interact virtually with each other. Simply put, it's a virtual world in which users can socialize, shop, perform different activities, and learn new things.

This is the trendy concept pushed by *Facebook*, which has made it its new mission and even its identity. The company defines it as a set of interconnected virtual spaces where users can share immersive experiences in real-time 3D. She sees it as "the next Internet." Other companies like *Nvidia*, *Roblox*, or *Epic Games* with *Fortnite* also talk about it as the future and the supporters of Blockchain and NFTs. They do not necessarily have quite the exact definition.

The term is a neologism coined by novelist Neal Stephenson for his book Snow Crash, published in 1992, from the word universe modified by the prefix "meta-" borrowed from Greek. No need to go into unimportant etymological considerations: the term serves as an alternative to "cyberspace," popularized by William Gibson, one of the creators of the cyberpunk genre. Stephenson describes it as the successor to the Internet, which takes the form of a 3D digital world in which users evolve as avatars.

The equivalent of a massively multiplayer role-playing game (MMORPG), so to speak, but used globally as the only source of entertainment or almost in a dystopian world. The author imagines a huge avenue ten thousand kilometers long, all along which there are various public or private spaces.

The rest is of little importance because the use of the word today has little to do with its origin, by the very admission of Mark Zuckerberg and the leaders of Meta.

Facebook CEO Mark Zuckerberg, who describes it as the future "Holy Grail of social interactions," has chosen to invest more vigorously in the Metaverse.

Here are two things to know about this concept that makes all digital heavyweights fantasize.

In recent months, the "Metaverse" has become one of the most fashionable words in the world of tech and video games.

History of Metaverse

Metaverse was founded by Eric Gu and Bobby Lee (CEO of China's first Bitcoin exchange BTC China) and is the only company in China that has received investment from the well-respected Sequoia Capital. Before founding Metaverse, Eric was a business development manager at *BTCChina*, a *blockchain* security engineer at *Circle* and *OKCoin*, and an investment manager at *Fenbushi Capital*.

Formerly a programmer at Tencent, Bobby is widely known as one of the co-founders of Litecoin and for his technical expertise in the Blockchain industry. He is also the founder of China's first Bitcoin club and co-founder of its first exchange.

Based in Shanghai, Metaverse's team already has members from several countries, including Singapore, Russia, Japan, and Hong Kong. The company has received investments from top VC firms, the most recent being a $4 million seed round led by Sequoia Capital China.

How it Works?

The Metaverse is a highly scaled and interoperable network of real-time provided 3D virtual worlds which can be experienced synchronously and persistently by an effectively unlimited number of users with an individual sense of presence and with continuity of data, such as identity, entitlements, history, objects, etc. The Metaverse has been a trendy topic and a major subject in 2021, with *Facebook* and *Microsoft* both predicting it to be the future of the Internet. How does the Metaverse work? When will it be a reality? Neal Stephenson, an author, is credited for coining the term "Metaverse" in "Snow Crash," his 1992 science fiction novel. He envisioned life-like avatars who met in realistic 3D buildings and other virtual reality environments. After that prediction, several developments have made bold claims on the way forward to a real Metaverse. This online virtual world integrates augmented reality, virtual reality, 3D holographic avatars, video, and other means of communication.

When the Metaverse expands, it will provide a hyper-real alternative world in which people coexist. Currently, glimmers of the Metaverse already exist in online game universes like Fortnite, *Minecraft*, and *Roblox*; and these companies behind the creation of these games have aspirations to be a part of the evolution of the Metaverse. The Metaverse is a combination of several elements of technology, including virtual reality, augmented reality, and video, where users live within a digital universe. Proponents of the Metaverse envision its users to work, play and stay connected with friends through concerts, conferences, and virtual trips around the world using VR headsets, and so on.

Popular Misconception about the Metaverse

It's also useful to analyze what the Metaverse is frequently compared to, albeit erroneously. Although these comparisons are most likely a component of the Metaverse, they are not the Metaverse themselves. The Metaverse, for instance, is not.

A Virtual World

Virtual worlds and games featuring AI-supported characters and those filled with "actual" individuals in real-time have existed for years. This is an artificial and fictitious universe, not a Metaverse with a single goal (a game).

A Virtual Space

Digital content experiences such as Second Life are frequently referred to as "proto-Metaverses" since they (A) lack game-like objectives or skill systems; (B) are persistent virtual hangouts; (C) provide near-synchronous content updates; (D) feature real people depicted by digital avatars. These, nevertheless, are insufficient characteristics for the Metaverse.

A Virtual Reality (VR)

VR is a method of immersing oneself in a virtual environment or place. A sense of presence in a digital environment isn't enough to constitute a Metaverse. It's the same as claiming to live in a prosperous metropolis because you can see and walk about it.

A Digital and Virtual Economy

Individual games such as World of Warcraft have had a functional economy for a long time, where real individuals sell virtual items for real money or do digital jobs for real money. Furthermore, platforms like Amazon's Mechanical Turk and technology like Bitcoin are built on the employment of individuals/businesses/computing power to execute virtual and digital jobs. We are already trading at scale for completely digital products for completely digital activities through completely digital markets.

A game Fortnite is a "game" with numerous Metaverse components, such as:

- Mashes together IP(B)
- Has a consistent identity that covers several closed platforms
- It is a portal to a diverse range of experiences, many of which are merely social
- Pays content producers, among other things.

Nevertheless, like with Ready Player One, it is still too limited in what it can accomplish, how far it can go, and what "work" can be done. While the Metaverse may have certain game-like aims, contain games, and use gamification, it is not a game in and of itself, nor is it focused on certain goals.

- A virtual Disneyland or theme park

Not only will there be an unlimited number of "attractions," but they will also not be "planned" or "coded" in the same way that Disneyland is, nor will they all be about pleasure or entertainment. Furthermore, the interaction distribution will have a very long tail.

• A new app store

Nobody needs a new method to launch applications, and doing so "in virtual reality" (for instance) would not unlock/enable the kinds of value promised by a successor Internet. The Metaverse differs significantly from today's Internet/mobile design, paradigms, and goals.

• A new UGC platform

The Metaverse will be a location where true empires are formed. The Metaverse isn't just another *YouTube*- or *Facebook*-like network where numerous people may "produce," "share," and "monetize" content, with the most famous content accounting for only a small portion of total consumption. Like with the web, a dozen or more platforms are likely to control a substantial portion of user time, experiences, content, and so on.

Building the Metaverse

The Metaverse will necessitate a plethora of new technologies, protocols, enterprises, breakthroughs, and discoveries to function. And it won't appear out of nowhere; there won't be a clear "Before Metaverse" and "After Metaverse" distinction. Rather, it will emerge gradually over time as various goods, services, and capabilities connect and meld. However, it's useful to consider three key factors that must be in place.

(One way I try to think about these three categories procedurally is through the *Book of Genesis*—first, establish the underlying universe ("concurrency infrastructure"), then describe its laws of physics and regulations ("standards and protocols"), and finally, one must fill it with worthwhile content ("Content") that evolves and iterates against selection pressures. In other words, God does not make and design the universe as if it were a small model but rather allows one to expand across a mostly empty tableau, etc.)

The Principle of Creating a Virtual World

The advantages are apparent, but businesses must keep several crucial concepts in mind to build the most powerful and interactive Metaverse experiences that will keep virtual users returning for more.

Design Unimaginable World

Don't simply design, reimagine. When it comes to creating an event or an experience, you have total creative freedom to create wholly unique and fantastical worlds.

Technology enables us to do the unthinkable—this is an opportunity to completely reimagine a setting rather than simply reproduce it and provide your audience with something unique that they would never be able to imagine in real life. Real-time audience participation, experimenting with dimensions, and displaying hypnotic images may all be used to improve the virtual world and create a really fantastic, unimaginable experience.

Do Not Neglect Emotion

Give your viewers the warm and fuzzies. The power to elicit emotion is at the heart of any effective experience, virtual or physical.

It is critical to generate compelling content that excites users and elicits an emotional response. Consider, for instance, a virtual concert. It is crucial to replicate the mutual flow of energy between stage performance and the crowd, which is achievable owing to immersive sound and stunning visual effects. An interactive Metaverse experience needs to stir up these human emotions, producing an environment that has a lasting impact on the users, much as a rush of adrenaline to the brain when you are among an agitated crowd.

Participation

Put the consumer in charge. Getting hands-on with a business is one of the most compelling brand experiences, and virtual environments should be no different. The audience is transformed into a user when they take an active role, which assures long-term influence.

Brand lovers are expressing an increasing demand for decentralization in a world where anybody can now be a creator. This can be facilitated by virtual worlds that provide real-time, AI-driven, or customized interactions that enable participants to control events in their world. We may allow audiences to influence events within a virtual business environment via interactive streams, aggregated comments, and click control—whether it be activating VFX, initiating audio, or even leaving their footprint on the environment.

Create Scarcity

This drives supply and demand. Even though Metaverse removes capacity limits and enables large masses to be reached, there are benefits to keeping particular activations limited to develop a feeling of uniqueness for visitors who are the "only ones" who enjoy that specific experience.

In the physical world, scarcity will always correspond to perceived worth. This impact must be produced in the virtual space—by restricting sign-ups, scheduling viewing times, or utilizing a blockchain system and ultimately boosting demand.

Continue to Evolve

Keep the relationship going. Businesses may not only own the design and customize the construction, but they can also collect data to monitor interaction and fine-tune the whole experience by creating their Metaverses. The greatest virtual brand environments will use these data to create fresh, inventive, and dynamic content to keep consumers interested and returning for more.

The complexity of traversing multi-platform, virtual user experiences will also be seen as the new CX frontier by intuitive businesses, who will adjust to purchase funnels that span both virtual and physical worlds to improve the experience even more.

Concurrency Infrastructure

At its most fundamental level, the technology does not yet exist to support hundreds, much alone millions, of individuals sharing a synchronized experience. Consider the Marshmello concert in Fortnite in 2019. Eleven million people in total watched the event in real-time. They did not, however, do so in tandem. There were actually over 100,000 Marshmello concert instances, each of which was slightly out of rhythm and limited to 100 players. Epic can certainly handle more today, but not into the hundreds of thousands, let alone millions.

Not only does the Metaverse require an infrastructure that does not already exist, but the Internet was never designed for anything remotely similar to this. It was, after all, created to transfer files from one machine to another. So, most of the Internet's underlying systems are based on a single server communicating with another server or an end-user device. This model is still in use today. Although billions of individuals are on today's Facebook, each user has a connection to the Facebook server and does not share it with anybody else. As a result, when you access content from another user, you're just getting the most up-to-date information from Facebook. Text conversations were the first pseudo-synchronous applications, but you're still feeding primarily static data to a server and pulling the most up-to-date data from it when/where/how/as needed. The Internet was never intended for persistent (as opposed to continuous) communication, let alone persistent communication synchronized in real-time with many other people.

The Metaverse requires something more akin to video games and video conferencing to function. These encounters operate because of persistent connections that keep each other up to date in real-time and with a level of precision that other programs don't require. However, they rarely have many concurrent users: most video chat systems have a limit of a few individuals, and after you reach 50, you'll have to "live stream" a broadcast to your viewers instead of sharing a two-way connection. These encounters don't have to be real, and they certainly aren't.

To that end, one of the reasons the battle royale genre has only lately been popular in video games is that it is only now possible to play live with so many other players. Although some of the most popular games, such as Second Life and Warcraft, have been around for more than two decades, they essentially faked the experience by "sharding" and dividing gamers into distinct "worlds" and servers. For example, Eve Online can have over 100,000 players "in the same game," although they are spread across multiple universes (i.e., server nodes). As a result, a player only sees or interacts with a small number of other players at any given time. Furthermore, getting to another galaxy necessitates quitting one server and loading another (which the game manages to "conceal" narratively by forcing players to travel at light speed to span the immensity of space). And if/when Eve Online reached battles with hundreds of players, the system ground to a halt. And it worked because the game's gameplay dynamic was primarily built on large-scale, pre-planned ship-based conflict. These slowdowns would have rendered the game unplayable if it had been a "fast-twitch" game like Rocket League or Call of Duty.

Like the appropriately called Improbable, several companies are working hard to overcome this challenge. However, this massive computational issue goes against the Internet's core design/intent.

Standards, Protocol, and Their Adoption

Standards and protocols for visual presentation, communications, graphics, file loading, data, and so on make the Internet work as we know it today. This includes everything from well-known brands to obscure ones. The WebSocket protocol, which underpins almost every form of real-time communication between a browser and other servers on the Internet, assigns GIF filetypes to it.

S&Ps will need to be even broader, more complex, and resilient in the Metaverse. Furthermore, because interoperability and live synchronous experiences are important, we'll have to prune some existing standards and "standardize" around a smaller set of standards per function—for example, today's image file formats include.GIF,.JPEG,.PNG,.BMP, TIFF, WEBP, and others. Even though today's web is based on open standards, much of it is closed and proprietary. Amazon, Facebook, and Google all use similar technologies, but they're not designed to work together like Ford's wheels aren't designed to fit into a GM chassis. Furthermore, these businesses are adamant about not integrating their systems or sharing their data. Such actions may increase the "digital economy's" overall value, but they also weaken their hyper-valuable network effects, making it easier for users to move their digital lives elsewhere.

This will be extremely difficult and time-consuming. The more valuable and interoperable the Metaverse becomes, the more difficult it will be to reach an industry-wide agreement on issues like data security, data persistence, forward-compatible code evolution, and transactions. Furthermore, the Metaverse will require entirely new rules for censorship, communication control, regulatory enforcement, tax reporting, the prevention of online radicalization, and a slew of other issues with which we're still grappling today.

While standard-setting usually entails face-to-face meetings, negotiations, and debates, the Metaverse's standards will not be established in advance. Meetings and opinions change on an ad hoc basis in the standard process, which is much messier and more organic.

Consider SimCity as a Meta parallel for the Metaverse. Ideally, the "Mayor" (i.e., player) would design their mega-metropolis first, then construct it from the ground up. However, you cannot simply "create" a 10MM-person metropolis in the game, as you cannot in real life. You begin by focusing on a tiny town and optimizing it (e.g., where the roads are, schools are, utility capacity, etc.) You build around this town as it grows, occasionally but prudently demolishing and replacing "old" areas, sometimes only if/when a problem (lack of power) or calamity strikes (a fire). However, unlike SimCity, there will be multiple mayors rather than just one, and their ambitions and incentives will frequently clash.

We do not know exactly what the Metaverse will require, much less how, when, or through which applications and groups existing standards will be transferred. As a result, it's crucial to evaluate how the Metaverse develops rather than just the technology standard it follows.

The Experience

Consumers and businesses will not adopt a would-be proto-Metaverse simply because it is offered, just as Metaverse standards cannot be "announced."

Take a look at reality. Making a mall large enough to accommodate a hundred thousand people or a hundred stores does not guarantee to attract a single customer or brand. To meet current civilian and commercial demands, "town squares" form spontaneously around existing infrastructure and activities. In the end, any gathering spot—whether it's a bar, basement, park, museum, or merry-go-round—is visited because of who or what is already present, not because it's a destination in and of itself.

The same can be said for digital encounters. Facebook, the world's largest social network, succeeded not because it declared itself a "social network" but rather because it began as a college hot-or-not, evolved into a digital yearbook, and became a photo-sharing messaging service. The Metaverse, like Facebook, must be "populated," not simply "populate," and this population must then fill up the gaps in this digital environment with activities to do and stuff to consume.

This is why thinking about Fortnite as a video game or an interactive experience is too limited and too quick. Fortnite began as a game, but it swiftly morphed into a social media platform. From the 1970s until the 2010s, teenagers would come home and speak on the phone for three hours. They now chat about Fortnite with their buddies, but not about Fortnite. Instead, they discuss school, movies, sports, the news, boys and girls, and other topics. After all, Fortnite does not have a story or an IP; the plot revolves around what happens on the island and who is present.

In addition, Fortnite is quickly becoming a platform for other businesses, IP, and stories to express themselves. This includes, most notably, last year's live Marshmello show. However, since then, the number of similar examples has exploded. In December 2019, as part of a larger in-game audience-interactive event that included a live mocap interview with director J.J. Abrams, *Star Wars: The Rise of Skywalker* released a clip from the highly anticipated film only in Fortnite. Furthermore, this event was mentioned openly in the film's opening scenes. Weezer created a special island where fans may get an early listen to their new album (while dancing with other "players"). In addition, Fortnite has created various themed "limited-time modes" based on *Nike's Air Jordan* and *Lionsgate's John Wick* film franchise. In other situations, these "LTMs" turn a section of the Fortnite map into a mini-virtual world that, when accessed, modifies the game's visuals, items, and playstyle to mimic something else. This has included the *Borderlands* universe, *Gotham*, *Batman*'s hometown, and the old west.

As a result, Fortnite is one of the few locations where *Marvel* and *DC*'s IP collide. You may practically dress up like a Marvel character and converse with others wearing officially approved NFL uniforms in Gotham City. This is the first time something like this has happened. It will, nevertheless, be crucial to the Metaverse.

More broadly, Fortnite has spawned a new sub-economy where "players" can create (and monetize) their content. Digital clothing ("skins") or dances ("emotes") are examples of this. It has, however, quickly grown to include all new games and experiences that use Fortnite's engine, assets, and aesthetics. There's something for everyone, from basic treasure hunts to immersive mash-ups of the Brothers Grimm with parkour culture to a 10-hour sci-fi epic spanning several realms and timelines. Fortnite's Creative Mode, in fact, already feels like a pre-Metaverse. A player enters a game-like lobby and selects from thousands of "doors" (i.e., space-time rifts) that transport them to one of the thousands of distinct worlds with up to 99 other players.

This relates to the game's long-term ambition, which creative director Donald Mustard is becoming increasingly clear about.

Who is Building the Metaverse?

Ice Stephen

Ice Stephenson (Paperback) Neal Stephenson coined the term *Metaverse* in his 1992 science fiction novel Snow Crash. The story unfolds in the future in Los Angeles after a global economic crisis toppled the United States government, dividing the country into several corporate-led states. Stephenson's Metaverse is an excellent multiplayer game, combining the unpopular truth of the taxpayers we see, the virtual reality, and the Internet, all connected and confidential. In this terrible situation, the criminal tries to plant a virus that will give him the power to control others. Snow Crash has inspired many tech leaders today, such as Epic Games CEO Tim Sweeney, Google Maps founder, Second Life Creator Philip Rosendale, and Xbox creator J Allard.

Meta Platforms, Known by Facebook

In a Facebook video last month, Meta Platforms CEO Mark Zuckerberg described his Metaverse vision as a 3-dimensional online version "that can bring a deeper sense of presence," where all the visible worlds are linked to the fact that user assets and information will interact with all countries regardless of their owners, whether Meta, Microsoft, or Epic Games, for example. Other important aspects of his view include the truth, the truth that the taxpayers may not like, and the creators. Users can access Metaverse with VR headsets, AR glasses, and other available devices. View it as a "mobile internet connection," where users will be able to work, play, shop, and visit friends and family. Digital asset ownership, now enabled by blockchain NFTs, will be important to the metering—and users will be able to bring their NFT, whether digital bike or batman costume, to all other forums and domains, regardless of their owner. Meta works on a few other projects within the Metaverse to create that sense of presence—new video games, avatars that can visualize your emotions and facial expressions, and digital rendering of real-world locations as the user move things about.

Microsoft

In an interview with Harvard Business Review, Satya Nadella states that the Metaverse will emerge gradually as the digital world becomes more physical and the physical world becomes more and more digital. His vision focuses on productive work, such as 3D immersion sessions, hologram avatars, and office spaces where colleagues on the web can come in to engage. According to the Washington Post, Nadella said the company is working to create a "business environment" and "put in a different place" to do so.

NVIDIA

NVIDIA embraces the concept of Metaverse as a reality widely shared on Facebook and Microsoft. Jensen Huang, CEO of NVIDIA, says NVIDIA's contribution to developing the universal project, creating and integrating digital gaps and virtual reality simulations—several businesses are already using the universe. In an interview with Hueng and Venturebeat, he says that the Omniverse simulation trains robots and explores ways to increase efficiency.

Apple

Apple CEO Tim Cook doesn't like to use different names. Many viewers still regard his company as a leader in the industry because of its focus and development of unpopular virtual reality products with the taxpayers we see. Apple is interested in creating more self-awareness using the unpopular reality of taxpayers we see. It recently unveiled new developer tools, including AR, and opened the exhibition of unpopular taxpayer art that we see in Istanbul. This principle—the increase in immersion and the integration of digital and physical realities—is in full harmony with the efforts of others to shape the Metaverse. According to Barron, Ming-Chi Kuo, an influential Apple analyst, expects Apple's first Metaverse hardware, AR glasses, to be ready by the end of 2022.

Digital Investors

Cryptocurrency enthusiasts view the growing Metaverse as a sign that it is time to invest in cryptocurrencies—some say the Metaverse would not be possible without it. In Coindesk's a crypto guide to the Metaverse, the author states that because cryptocurrency solves many of the current Metaverses problems, it will be important to Zuckerberg's future Metaverse thought. The author referred to Decentraland and The Sandbox, two major multiplayer games where cryptocurrency is used as real money to support the claim.

They both call themselves Metaverses. Instead of investing in cryptocurrency, others have invested in digital real estate. The best-selling international price in Decentraland was over $ 500,000. The theory is that as more people enter the Metaverse, these digital spaces will increase.

According to Mark Zukerberg

The "Metaverse" years are upon us—or at least that's what Mark Zuckerberg would like us to believe. To this afternoon, known as Facebook, the Meta manager spent more than an hour Thursday with viewers at the annual Connect Engineers conference, so he does not expect the next version of the Internet to work and the hardware and software. That will get us there. The Meta event was ambitious, especially with the standards of the developer conference, which means we have a lot to offer. Here's what you need to know about Meta programs and what that means for people who may (eventually) use them.

2.

Meaning of VR (Virtual Reality), Terminology, and Concepts

Virtual reality, or VR, enables individuals to be immersed in a simulated world. This is generally (but not always) accomplished by using head-mounted hardware that records an individual's motions. A screen (or two display panels, one for each eye) is enclosed in a frame (or headset) that is fastened or attached to your head. A pair of lenses is generally attached between the panels and your eyes, shutting off the outside world to make it look like what you see via the headset is your whole environment.

Importantly, all headsets monitor your movement and appropriately modify the picture you see. Some headsets, on the other hand, monitor movement more than others. Virtual reality experiences usually provide a way for you to manipulate or select items inside the simulated world. In certain cases, you'll have a controller in each hand to control different parts of the game. In certain situations, the controllers provide virtual depictions of your hands that may be used to influence the environment and items inside it in ways that are comparable to those found in the real world.

Types of VR

There are different types of virtual reality equipment on the market. These are divided into three categories: standalone VR, PC VR, and console VR.

Standalone VR

Any VR headset that functions fully independently of any other equipment or technology is a standalone VR. The whole experience is controlled by the headset you wear on your head and does not need any other devices.

The finest example is Oculus Quest 2, which provides stripped-down versions of PC VR games in a compact and independent device that doesn't need additional hardware. Everything you require to enjoy virtual reality comes with standalone VR like the Quest. However, certain games require you to hold the Oculus Touch controllers in your hands to participate.

PC VR

Any VR that needs a continuous connection to a close PC is referred to as PC VR. The PC in question will, at the same time, need to have sufficient specifications to support VR. The Oculus Rift S, Valve Index, HTC Vive, Pimax, and Windows Mixed Reality headsets such as the HP Reverb G2 and Samsung Odyssey+ are just a few examples of PC VR headsets.

The benefit of PC VR is that it can give considerably greater graphics quality than standalone VR due to the hefty PC specifications. Several wireless PC VR alternatives usually need additional hardware linked to the PC, and a battery pack carried somewhere on the body. In wired PC VR, you must handle a cord that connects the headset to the computer. The cord can be a constant reminder that you can get tangled when you spin around too much. Thus it may give less flexibility than a wireless solo headphone.

PC VR may rapidly become a costly alternative if you don't already have a gaming PC that fulfills the minimum specifications.

Console VR

For the Nintendo Switch, there are just two virtual reality headsets available: PlayStation VR and Nintendo Labo VR.

PlayStation VR is a PS4 or PS5 VR add-on system. The PSVR headset is an add-on that attaches to the PS4 and, like PC VR, needs a continual tethered connection to the system to work. It tracks with the PS camera, which comes with the headset and is linked to the system. Thanks to backward compatibility, the system equally works on PS5, with minor aesthetic and performance enhancements. To enjoy some PS VR games, you'll need to buy PlayStation Move controllers separately.

Nintendo Labo VR for the Nintendo Switch is a cardboard VR headset shell built and inserted into the original full-size Nintendo Switch, enabling you to play selected games in virtual reality. Because there is no VR headset strap, you must keep the headset to the face. Most of the Labo VR experiences are, to put it bluntly, disappointing and not worthy of your time.

Major Players

Oculus VR

In 2016, Facebook paid a staggering $2 billion for the firm. John Carmack, the creator of id Software's 'Doom,' was also part of the team. Carmack later departed Oculus due to legal conflicts with ZeniMax, but the company's creative engines have not slowed down.

They have unveiled the 'Oculus Quest,' a standalone headset. Oculus is not just a name to remember; it is *the* name to remember when it comes to virtual reality firms.

Google

Yes, Google is among the major players in the virtual reality space. Gone are the days of searching for things like "how to tie a tie" and "pie recipes." Google is currently experimenting with virtual reality. It was a fantastic move for them to appear on the scene. They unveiled the 'Google Cardboard,' a $15 cardboard virtual reality headset.

It was created to be worn over your phone or device and engage with various apps and games. It is easy to imagine Google continuing to investigate the field of virtual reality and even branching out with a greater presence.

HTC Vive

The HTC Vive is a hybrid of The Matrix, Ready Player One, and the cyberpunk genre. This headset was introduced by HTC for $799, distinguishing experts from beginners. It comes with a headset, room sensors, and portable gadgets. One of the things to admire about HTC is that they collaborated on this initiative with Valve to make it more powerful.

This, I am sure, applies to a lot of VR Steam games. They received such a good mark not just because of the technology they provide but also because this VR business is teaming up with a firm to change the VR sector with tanklesslab.

You will be walking around the area with your headset, not just sitting in a lonely spot.

Unity

Unity is a well-known developer in the area of 3D animation. Therefore, it is no coincidence that they have produced VR-ready software. Most interesting virtual reality content has gone through the Unity 3D engine, making Unity an important VR and a leading VR firm.

One look at their site reveals an outstanding portfolio of achievements and content to use in your creative efforts.

Microsoft

Microsoft is another famous name and another potential virtual reality heavy hitter. Microsoft joins the list, demonstrating that virtual reality companies can be large, established businesses. Microsoft is putting a lot of emphasis on VR and AR. The HoloLens is now available from Microsoft, and it enables not just games but also real applications.

Samsung

Many of the names we are familiar with from our daily lives appear here. They do so because, fortunately for them, they already have a technological head start, and Samsung is no newcomer to technology, including virtual reality. Samsung made a good move by collaborating with Oculus to create the Samsung hardware, a virtual reality headset.

It is more affordable, making it a more appealing choice for casual users. It is a VR headset for your smartphone, and you just slip the smartphone into the headset. Samsung has a lot of promise in this area, and I believe they can even start making games for their projects.

Magic Leap

Magic Leap is a startup business that is a shift from the list of virtual reality companies. Magic Leap is concentrating on augmented reality rather than complete virtual reality. They employ 3D elements to populate the scene surrounding the user, which I find fascinating. This offers the possibility of future development since they may become the market leader in augmented reality.

Magic Leap is marketing its 'Magic Leap One' with the phrase 'Free Your Mind.' They are just selling the creator version of the VR, so hopefully, the applications and games it uses will be completely functional by the time it hits the market.

WorldViz

WorldViz mostly creates virtual reality programs for schools. They also provide programs for architectural and safety training. Consider how beneficial it will be for architects to explore the many models they create rather than relying just on drawings and models. It would alter the way buildings are constructed, which is why more firms like WorldViz are needed.

Snap Inc.

Snap is the business that created Snapchat, so you know these folks are on top of their game in terms of technology. They are experts at incorporating AR into their technology, demonstrating that they can be a top VR firm. Every day, you may use your phone camera to explore different parts of A for free.

You may use 3D models to enhance pictures or videos of yourself or your pals. This technology is making its way into the public without requiring hundreds of thousands in development and release.

Wevr

Wevr is largely driven by user-generated content. This technology is placed in the hands of ordinary people, which is quite appealing. With their initiative, they plan to make a mainstream virtual reality, the "YouTube of VR." Users will upload their content for people to enjoy in VR, making the process smooth and straightforward.

Firsthand Technology

This firm is promising due to its mission. It is a noble cause that VR firms should prioritize. They are primarily concerned with healthcare. The programs assist individuals who suffer from anxiety in reducing their symptoms. Imagine you have had a particularly stressful day, and your nerves are frayed. You put on your VR headset and begin to relax.

Because their product is dependent on heart rate, they advise consumers to relax while using it. This passes a significant threshold in VR; the company and product genuinely care about the user.

NextVR

NextVR is the virtual reality company for sports enthusiasts out there. It broadcasts sporting events, allowing you to feel closer to your desired sport.

Imagine being able to feel like you are at a Philadelphia Eagles game, such as their recent Super Bowl triumph against the New England Patriots, without having to pay the exorbitant ticket fees.

Nvidia

We are all familiar with Nvidia's legendary commitment to producing top-of-the-line graphics cards that allow you to enjoy your favorite games in the highest possible resolution. Nvidia, like Unity, provides virtual reality technology for creators and other virtual reality firms to employ to develop better apps and games.

They enable a clearer and fluid resolution when a participant or user is in the program or application, resulting in a more realistic experience for all parties involved.

Prenav

Drones, like virtual reality, have become a fixture in technology. Prenav provides drones with more sophisticated software to explore structures better than basic Google Maps or other GPS technologies. These drones keep the cell service working efficiently by building 3D models of mobile phone towers for drone inspection.

The significance of everything functioning is the primary reason for their high ranking on the best VR businesses list.

Osterhout Design Group

Osterhout brings the Watch Dogs gaming series to life for individuals who have played it. Their headset—or, more precisely, their glasses—provides you with useful information about what you are looking at. In comparison to Google Glass, Osterhout's technology may provide you with biographies of persons you are looking at and other valuable features.

Imagine receiving a piece of furniture and being completely baffled as to how to put it together correctly. Osterhout assists you in putting it all together, removing the difficulties that come with some elements of daily living. That is such an essential objective for a VR firm.

Application

Gaming

Entertainment and Culture

In recent years, mainly due to the new wave of commercial viewers initiated by Oculus Rift, the general public's interest in virtual reality has been rekindled, especially in the field of gaming and, in general, entertainment. In addition, other activities and areas that are by nature more education-oriented have opened up to the novelty of VR to try to attract the younger audience, usually more attentive to technological innovations; thus, we can find virtual museums and theme parks, interactive theatrical performances and more.

The main driver that unites these new experiences is the public's sense of immersion and involvement, which is no longer a passive part but can have, if desired, an active role. The direct interaction that new virtual and augmented reality technologies allow becomes a means to communicate information in new and interesting ways.

Think, for example, of a museum that exhibits computer reconstructions of tools and machinery from the past; by wearing VR visors, the public can observe the 3D reconstruction from different angles, as if they had the original object in front of them; moreover, since it is a 3D model, it is possible to allow the public to interact with the reproduced tool or machinery, increasing, on one hand, the involvement and, on the other hand, the learning aspect.

Examples similar to that of the machinery reproduced in 3D are the computer reconstructions of architectural elements, or even entire historic and ancient buildings no longer existing, as destroyed by time or natural or manufactured disasters. Through VR, it becomes possible to bring to life what no longer exists and admire it with your own eyes in a more natural way than, for example, watching a reproduction on television. In addition, thanks to the tracking system of modern VR viewers, you can move freely in space to observe the reconstruction from different points of view.

Architecture

When buying a house, redoing the kitchen or the bathroom, or more generally when renovating a house, it is becoming increasingly common for the client to have the opportunity to see in advance the final result of the work, faithfully reconstructed in 3D, sometimes even in immersive mode, i.e., using a Virtual Reality helmet.

This practice, called architectural visualization, allows the client to evaluate the result and the architect to verify the validity of his project. In addition, since virtual reconstruction is much less expensive and time-consuming, it is possible to make changes, even frequent ones, to the original project; by evaluating different factors and aspects of the project before the actual construction, design errors are greatly reduced.

Visiting virtually what is to be built allows, finally, to evaluate first-hand the dimensions and spaces, which would otherwise be difficult to estimate from a technical drawing for a common person.

Another area where the ability to observe the final product before it is actually created can be extremely useful is in the construction of large buildings and skyscrapers. For example, doing a simulation of the construction can help optimize the construction process and prevent any critical issues before they arise, thus reducing the time it takes to create the building.

Production and Marketing

In the manufacturing phase of a product, processes based on new technologies such as virtual reality can replace traditional processes that are deemed more expensive, unsafe, and inefficient.

For example, before a product reaches its final form, it undergoes a series of changes, evolving from its initial idea to become the product you find on the shelves of your local store. The prototyping process can be extremely costly and time-consuming, though necessary to achieve the quality standards required by the customer. In all, VR can be a very cost-effective and time-saving solution for prototyping, which is why it is gradually replacing the traditional prototyping process wherever possible.

In other cases, cost and production time are not the only determining factors in choosing the type of process to adopt, but also the safety of engineers, workers, and company and/or institution personnel in general. Being by nature a simulated environment, VR eliminates many of the risks associated with the production of potentially dangerous machinery. In fact, a worker can be trained in the virtual environment before coming into contact with the actual machinery; on the other hand, during production, certain operations can be performed in the virtual environment to prevent any unforeseen complications before being performed on the real machinery.

Military

In the military field, virtual reality is used as a tool for training; it is particularly useful for training soldiers to deal appropriately with conflict scenarios and unusual and dangerous situations without running the risk of being seriously injured or dying.

Compared to traditional training methods, VR simulation is less expensive and does not put soldiers' safety at risk. Devices commonly used by soldiers in training sessions are head-mounted displays and data gloves, both of which are tracked to allow interaction with virtual objects.

1. Typical training activities include:

2. Training of medical personnel.

3. Simulation of armed conflict.

4. Flight simulation for pilots.

5. Driving simulation.

Virtual reality can help in treating symptoms caused by post-traumatic stress disorder (PTSD), a condition shared by many veterans who have suffered from traumas or particularly difficult psychological conditions on the battlefield. To teach patients how to manage their PTSD symptoms, they are subjected to situations that gradually trigger the disorder so that they can learn to control it in an environment they perceive to be safe.

Outside Gaming

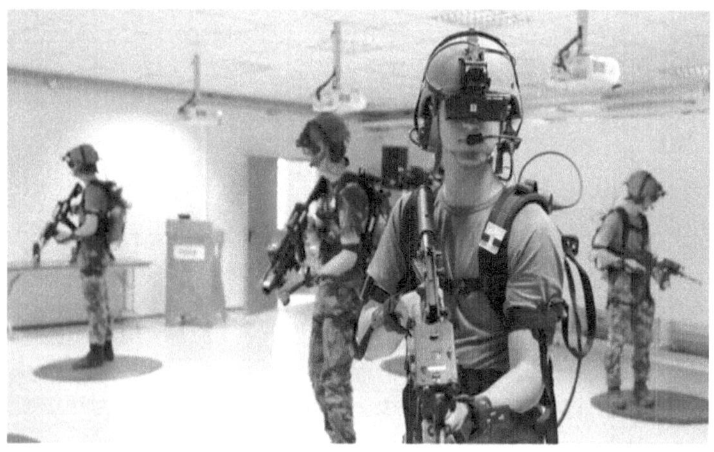

Virtual Reality in the Military

Both the British and American militaries have used virtual reality in their training because it allows them to do a wide range of impersonations. VR is used by all military departments, including the navy, army, air force, marines, and coast guard. VR may efficiently transfer a learner into different scenarios, places, and surroundings to facilitate teaching.

Military applications include aviation, vehicle, warfare simulations, medic training, and creating a virtual boot camp.

The technology provides a completely engaging experience with sights and music that can safely replicate dangerous training exercises for preparing and training soldiers while preventing exposure to danger until they are battle-ready. Additionally, while on the battlefield, the technology may educate soldiers in interacting with residents or foreign journalists.

Another use of virtual reality is for treating Post-Traumatic Stress Disorder (PTSD), which troops were returning from battle frequently experience. They require aid in readjusting to civilian life.

Virtual Reality in Education

In addition to teaching and learning situations, virtual reality is being used in the education industry. It enables students to converse with one another in a 3D space. Students could also go on virtual field trips to art galleries, take solar system tours, and go back through time to other eras.

VR can be particularly beneficial for kids with special needs. According to research, virtual reality might be a compelling platform for properly training and teaching social skills to children with autistic problems. For example, *Floreo*, a technology firm, created VR scenarios that allow kids to acquire and practice eye contact, pointing, and forming social relationships.

Virtual Reality in Sports

VR has been slowly reshaping the sports business for all of its players. Coaches and athletes may use this technology to successfully train across various sports since they will be able to watch and experience certain scenarios again, improving their performance each time.

Virtual reality is now being used as a training tool to assist in evaluating sports performance and skill. It has also been shown to help wounded players improve their mental skills by enabling them to experience game conditions digitally.

Similarly, the technology is being used to enhance the viewer's experience when viewing a sporting event. Many broadcasters have begun live streaming using VR. They are planning to offer virtual tickets for live sporting events, allowing fans worldwide to attend any sporting event. This also allows individuals who cannot afford to spend money to attend live sporting events and feel involved since they can have a comparable experience from the comfort of their own homes at no cost or a lower cost.

Virtual Reality in Mental Health

Virtual reality technology is being used to treat PTSD. An individual is immersed in a simulation of a traumatic experience using VRTD (Virtual Reality Exposure Therapy) to help the individual overcome the situation and begin to recover.

It is also used to treat emotions like anxiety, sadness, and phobias. Several individuals with anxiety, for example, have identified meditation using virtual reality as a beneficial way to handle stress sensitivity and improve coping skills. Virtual reality technology can provide a safe atmosphere for patients to confront the things they are afraid of while being protected and secure.

Virtual Reality in the Fashion Industry

One area of virtual reality's use case in the fashion business that has received far less attention is its use in the fashion industry. Virtual simulations of shop settings, for example, may be a very efficient way for merchants to practice building their signage and product exhibits without having to commit to the actual build.

Similarly, adequate efforts and money may be allocated to creating a shop layout. Tommy Hilfiger, Coach, and Gap are just a few of the well-known businesses that have begun to incorporate virtual reality into their operations. These companies are using virtual reality to provide customers with a 360-degree view of fashion events and allow them to try on items virtually.

3.

AUGMENTED REALITY

Augmented Reality (AR) and Evolution

Augmented reality is a technique of exploring the natural world through a device like your phone or camera or directly relating with the world without a digital intermediary. The real world is augmented into a real-world visual with inputs generated from the computer like audio, graphics, or videos. Augmented reality is different from Virtual Reality because augmented reality adds something to an already existing world. It is not creating an entirely new world for its users from scratch, and it just builds upon what is living already in the world. The natural and digital worlds have no route to interact or even respond. But augmented reality propounders created a mixed reality hybrid where interaction between the augmented digital world and the real world exists.

Evolution and Types of AR

AR's evolution may be broken down into three distinct stages:

Attention Stage

Ivan Sutherland, a computer scientist, invented the first *Augmented* reality technology in 1968. He was also referred to as the "Father of Graphics." Sutherland invented the first augmented reality (AR) head-mounted display.

In addition, the year saw the development of one of the first AR applications for commercial usage. The creation of AR applications was done for the aim of advertising. AR was utilized in advertising by Munich-based German agencies. They developed a printed commercial for BMW Mini. The same appeared on the screen when the automobile model was held in front of the camera.

In 2008, the first commercial augmented reality application was released. It was created for advertising reasons by Munich-based German firms. They created a printed magazine ad for a BMW Mini model that appeared on the screen when held in front of a computer's camera. Eventually, major corporations like Disney, National Geographic, and Coca-Cola began to use Augmented reality and Virtual reality technologies.

Augmented reality has been used for a variety of applications. For example, National Geographic used AR to depict ancient creatures walking around a showroom rather than in a cage!

Coca-Cola employed augmented reality to raise awareness of the melting ice's environmental impact. Individuals had only heard of melting ice until that point, but Augmented reality allowed them to experience it in a shopping mall. In terms of increasing environmental issues, the effect would have been doubled.

Disney employed augmented reality to display cartoon characters on a huge screen and have them talk to people. It would have been fascinating to see!

Trial Stage

The second stage of Augmented reality development began when digital items were modeled. Thanks to the simulation, the products may connect with the physical world and move in real-time.

In the 2010s, the product classes that used Augmented reality were jewelry and watches. Throughout this era, the usage of AR for product "try on" became widespread. Apple Watch also provided a virtual try-on for bringing its watch to life.

Usage Stage

Augmented reality was being used in a wider number of sectors at this point in its development. AR has progressed to the point that it is now one of the most useful technology for exploring geographic, historical, cultural, and environmental elements.

In the past, Augmented Reality was used to transport travelers. Evolution applications for the tourism sector were created during this stage of Augmented reality. For instance, the Museum of London's augmented reality application allows guests to feel and explore streets from 1000 years ago. The AR software also brings the artworks in the museum to life.

Likewise, museum applications allow visitors to learn more about great artworks by putting a description on their phone displays in real-time

Types of AR

The intriguing and simple thing about augmented reality is that anyone can use it. You don't have to be an internet guru before you can operate in the augmented reality world. The companies that created augmented reality are helping users derive fun in using augmented reality with the assistance of smartphones. That is why augmented reality is popular.

The types of augmented reality currently in the market are:

Marker-Based AR

The marker-based AR is also referred to as Recognition based AR or Image recognition. In this kind of Augmented Reality, you have access to more information about a particular object you want to find information about. So, augmented reality focuses on that object and gives you the necessary information you need. Marker-based Augmented Reality has a different use. It can discover the object directly in front of your camera. The user can look at the object with more detailed information from different angles. As you rotate your marker, the 3D imagery will also rotate.

The Superimposition Based Augmented Reality

- Augmented reality gives you a view of an object that is in focus.

- Superimposed-based Augmented objects will have an augmented view.

The Markerless Augmented Reality

It is also called location-based augmented reality because the features are readily available in smartphones to detect locations. Most people use this kind of app to give them directions while traveling. It also helps the individual look for interesting places in their most recent locations. You can read the data from your mobile phone's GPS, accelerometer, and digital compass while also finding out where the user is standing. You need to add information about the object's location on the screen using your device camera for this AR to work perfectly.

Projection Augmented Reality

Projection Augmented reality is the most accessible type of augmented reality. It is all about projecting a stream of light on a particular surface. The projection-based Augmented Reality is interactive and appealing when light is shown on a surface, and communication takes place by simply touching the projected surface with your hand. The projection-based augmented reality is mainly used to create the position, depth of an object, and orientation. Try out this tech to create virtual objects with more massive deployment.

Application of AR

AR is an advanced technology that aids businesses in enhancing consumer satisfaction. This section will concentrate on how AR may be used in the real world.

AR in Airport Navigation

Terminal 1, Terminal 2, Terminal. It is never simple to find your way around an airport. The Gatwick Airport passenger application was created utilizing augmented reality to help travelers navigate the airport. The software navigates travelers around the airport using over 2,000 Bluetooth beacons. This software might become one of the instruments for managing airport traffic flow.

AR in Manufacturing

Did you have any reservations about buying furniture? The color may be a mismatch with the walls. The furniture may appear tiny and large in the room, and several other flaws. Customers will benefit from the Ikea Place app, which is an augmented reality software. The software uses ARKit technology that enables users to scan a space to purchase furniture. By adding things to a new digital world, you may simply create the room and choose the color of the furniture that matches.

AR in Cosmetic Store

What color gloss should I wear if this one does not match my skin tone? Questions that are often asked when purchasing cosmetics. Cosmetics are an expensive purchase, and making the incorrect selection may result in money being wasted. *Sephora*, a cosmetics firm, is aware of this problem among its clients. As a result, the firm employs augmented reality technology to allow customers to sample items in a realistic setting before purchasing them!

AR in Healthcare

AR can also save lives! Yes, you read that correctly. Are you assuming that Augmented Reality will take the place of medications and procedures? No, definitely. In healthcare, augmented reality applications will give real-time data that may be utilized for patient diagnosis, therapy, and operation planning.

AR in Games

The concept behind augmented reality games and applications like *Pokémon Go* and *Snapchat* is now obvious. Is there any augmented reality game that comes to mind? By

2023, the market for augmented reality games is anticipated to reach $284.93 billion. Other than Pokémon, there are various games to choose from, like *Zombies Run, Ghostbuster World,* and *Kings of Pool.*

Education

AR has the potential to improve teaching and learning. Augmented reality can change textbooks and schools by converting previously static data and photos into immersive experiences. When you can use AR to dissect the strata of a volcano or plunge hundreds of miles into the Earth's crust, geology becomes a lot more interesting.

One of the most basic study aids, even flashcards may benefit from augmented reality. Augmented reality Flashcards Animal Alphabet is an application that uses augmented reality to assist young kids in learning the alphabet by bringing flashcards to life. When the duck from the "D is for a duck" card stands in front of you, the ABCs appear a lot easier and more enjoyable.

Sports and Entertainment

Even the manner we purchase Super Bowl tickets is evolving thanks to AR technology. StubHub launched a feature on its smartphone application for Super Bowl LII that allows ticket purchasers to view a virtual 3D model of US Bank Stadium and the surrounding region. This was not the firm's first time experimenting with augmented reality. *StubHub* formerly provided a "virtual view," which enabled people to experience the view from their seats before purchasing a ticket. StubHub saw its engagement increase in a year after releasing that option.

Many sports leagues have also used AR to improve their viewers' watching experience. The MLB's famous "At Bat" application now includes augmented reality capabilities that will enable users to see actual stats on every player, ball speed and distance covered, and other data by just pointing their phone at the field.

VR versus AR

The difference between augmented and virtual reality lies in the details: while VR creates the whole environment and the user is an anonymous part of it, AR includes real-life elements like objects and people. This doesn't mean that VR is not interactive; in fact, with features like 360° head tracking and hand gesture recognition, VR has become quite immersive.

Mixed and Extended Reality

Mixed Realities

Mixed reality is the blending of virtual realities with real-life environments. You can already see it happen on your smartphone or tablet with apps that add a digital effect to photos or videos, but that's just the start. Mixed reality delivers realistic, responsive, and interactive experiences by blending the virtual world into ours—you'll interact with it, connect to other people through it, and ultimately use it to transform the way you live.

"Enter Mixed Reality—a virtual world that is experienced in the real world by using devices such as VR headsets and AR glasses" (Kumar).

What is Extended Reality (ER)?

Extended reality allows employees to better understand objects, concepts, and processes by experiencing and seeing them. For instance, an MR app can allow illustrative anatomy training that shows organs layer by layer and displays how they work.

Pros and Cons of Extended Reality

For today's world, XR promises a new terrain of exploration and creativity. The potential of extended reality will only expand when new technology in haptic technology and cognitive control develops. The following are some of the advantages and disadvantages to consider right now:

Pros

Limitless Creativity

Visualizing and playing with 3D representations of items and designs allows for far more innovation.

Enhanced Collaboration

Unlike traditional video or voice conferencing, XR may bring individuals together with a far more realistic sense of immersion.

Improved Learning

In practically all contexts, having access to contextual information, demonstrations, and 3D representations enable improved and greater learning experiences.

Enhanced Productivity and Effectiveness

In an extended reality environment, workers can collaborate and work much more quickly and efficiently than they could if they were exchanging emails.

New Experiences

From digital events to holidays, extended reality technologies allow us to discover new experiences like never before.

Cons

Privacy Concerns

Many individuals are concerned about the privacy implications of digital technologies. Extended reality technology requires a large amount of data about the participant to function properly.

Integration Costs

High-quality extended reality experiences are costly, especially when considering headsets and haptic sensors. Not everyone has the financial means to participate.

Physical Safety

Entering a new virtual world or supplementing an old one can make it more difficult to be aware of potential threats.

4. ARTIFICIAL INTELLIGENCE

Why Does the Metaverse Need AI (Artificial Intelligence)?

Artificial intelligence (AI) functions are now much more common than you think. For example, 50% of respondents said that their companies use AI for at least one business function in a recent survey. In addition, according to the study, 40% of companies have adopted a strategic plan across the board.

AI now plays a big part in consumer-facing apps through facial recognition, natural language processing (NLP), quicker computation, and a variety of other behind-the-scenes operations. So it was just a matter of time before AI created brighter immersive worlds in augmented and virtual reality.

AI can unravel large volumes of data at lightning speed to develop insight and encouragement. Users can use AI for decision-making (because most company applications do) or combine AI with automation for low-touch activities. The Metaverse will combine augmented and virtual reality (AR/VR) with artificial intelligence and Blockchain to develop scalable and accurate virtual worlds.

AI Use Cases in the Metaverse

While VR worlds can technically survive without artificial intelligence, the combination of the two unlocks a whole new degree of authenticity. This could have an impact on the following five use cases:

Avatar Creation

The Metaverse revolves around its users, and the accuracy of your avatar will affect the quality of your and other participants' experiences. For example, an artificial intelligence system may analyze 2D user photos or 3D scans to create a realistic virtual reproduction.

It may plot various facial expressions, emotions, hairstyles and age-related features and make the avatar more vivid. Many companies have used AI to help build avatars for Metaverse, and Meta is working on its technology version.

Digital Humans

In the Metaverse, digital humans are the 3D counterparts of chatbots. They're not precise replicas of real people; instead, they're AI-enabled non-player personalities (NPCs) in a video game that react and respond to your actions in a virtual reality environment. Digital beings are created totally with AI technology and are vital to the Metaverse's ecosystem.

Numerous applications range from NPCs in games to automated assistants in virtual workplaces. Firms like Unreal Engine and Soul Machines have already invested in this direction.

Multilingual Accessibility

One of the most important uses of AI in digital humans is language processing. Artificial intelligence can help break down human languages like English, transform them into a machine-readable format, analyze them, develop a response, convert the results back to English and communicate them to the user. This whole thing happens in a fraction of a second, much like a real conversation. The best part is that, depending on the AI's training, the results might be translated into any language, allowing users from all over the world to access the Metaverse.

VR World Expansion

This is where artificial intelligence shines. AI machines learn from previous results and efforts to produce themselves when given historical data. With new input, human feedback, and strengthening machine learning, AI output will improve over time.

Eventually, AI will complete the task and provide almost as good results as humans. Companies like NVIDIA are training AI to develop an entire virtual world.

This breakthrough will play a role in the scalability of driving for Metaverse because the new world can be added without human intervention.

Intuitive Interfacing

AI can also help in human-computer interactions (HCI). When you wear a sophisticated AI VR headset, the sensor will be able to read and predict your electrical pattern and muscles to know precisely how you want to move into Metaverse. AI can help reinvent a taste of authentic touch in VR. It can also help in navigation that supports sound, so you can interact with virtual objects without using a hand controller.

Challenges Around AI in the Metaverse

It is important to remember that Metaverse is a new field of research and operations, and the implementation of AI can experience problems. For example, there may be questions around:

- **Ownership for AI-created content**: Who owns the rights to the content and virtual worlds created by AI, and who can benefit from them?

- **Deepfakes and user transparency**: How do you ensure that users know they interact with AI and no other humans? How do you prevent false and fraud?

- **Fair use of AI and ML**: Is it legal for users to employ AI/ML in Metaverse interactions? Can they, for example, is it possible to use AI to win games?

- **Right to use data for AI model training**: How can we educate AI for the Metaverse in an ethical manner? What are the processes for obtaining consent?

- **Accountability for AI bias**: If a digital human or similar AI algorithm displays bias, what is the possible recourse?

NFTs and their Role in the Metaverse

NFTs, or Non-Fungible Tokens, are a new type of crypto coin that can be used in games. NFTs are like Beanie Babies—each one is unique and has its own identity. This means you could collect NFTs from different sources and combine them to make a more valuable set.

Non-Fungible Tokens, or NFTs, are a new form of digital currency on the *Ethereum blockchain*. Unlike traditional cryptocurrencies such as *Bitcoin* and *Litecoin*, they have a unique token for each item they represent. This is to say that each token is different and cannot be substituted by another one. These tokens can represent any type of asset, including in-game items, land titles, real estate deeds or stocks, and shares. The technology has more applications, too—NFTs are also used to track unique animals such as tigers and pandas and assist in managing copyright law for creative works.

NFTs are essentially just a new name for the collectibles on the Ethereum Blockchain that we first got to know back in 2016. By extension, these tokens can be used to represent any kind of asset on the Ethereum Blockchain.

NFTs are the fastest growing asset class of this century, with many new assets being introduced every day. This book will go over the importance of NFTs and what to look for when trying to invest in this area.

It also goes over the different types of NFTs available to maximize your chances of finding a worthwhile investment: tokens, collectibles, digital goods, notary services, and derivatives. When deciding which type to invest in, you'll have to consider many factors, such as liquidity and future potential.

The Blockchain has been around for almost 10 years now and continues to expand with new assets being introduced every day. This will explain the importance of NFTs and go over the different types of NFTs available.

Non-fungible tokens are unique digital tokens designed to represent something specific. The most popular example of NFTs is CryptoKitties, which originally introduced the concept of digital assets you can collect and trade. Each token is very unique in its own way, meaning a CryptoKitty can have a low supply or be very rare.

The term NFT is used in the same way as collectible items such as Beanie Babies or Pokémon cards. There is a demand for these unique items, and some people spend a lot of money on them.

Role of NFTs in Metaverse

NFTs assist the Metaverse in achieving three of the seven fundamental criteria stated by Matt Ball in his Metaverse blog:

- **Mass Participatory Medium:** The Metaverse will be a hugely interactive technology that will run simultaneously and be driven by the creative effort of millions of individuals.

- **An Independent and Fair Economy:** The ideal future Metaverse will be based on an autonomous economy, which is open and decentralized, allowing for worldwide job opportunities.

- **Individual and Collective Agency:** The Metaverse will fluidly enable individual agency (i.e., you develop and produce an action) and also collective agency (i.e., people act collectively toward a common goal, such as in a mass movement), giving you a feeling of control over your Metaverse experiences.

Large firms will certainly impact the Metaverse by creating environments, entertainment experiences, games, and the system that supports them. On the other hand, the Metaverse will be a lot more diverse and creative media than we are used to. While social media networks make it possible for anybody to create content, they are quite limiting. They decide who has access to them, what content may be generated, and who benefits from the creators' efforts.

The Metaverse enables anybody to engage as an important member of the Metaverse and actively shape its appearance.

NFTs will be a crucial component of realizing that creative participative future. In the future Metaverse, NFTs can contribute to generating meaning and economic activity, and offer utility.

Contribution of NFT to Metaverse

Since NFTs are just digital information of physical objects, they can be used in a variety of ways. For example, NFTs are used to create different online identities that can be registered on the Metaverse platform; frequent traveler mileage cards;

artwork authentication; jewelry authentication; and other types of certification. Moreover, NFTs allow people to make a trade between themselves without ever needing to meet each other face-to-face.

NFTs could be composed of a variety of information types. For example, it could be pictures and texts, location data, or even low-resolution videos. The information types can be anything that can be displayed on the blockchain and shared amongst users.

The Metaverse has been talked about for a couple of decades now, but the discussion certainly intensified with the technology the blockchain and NFTs brought. But what is it? Is it already happening? Will it happen in the near future? Will it happen at all?

Well, that depends on your own definition of the Metaverse. Something is certainly happening in the blockchain. Ask the art world, people who are making bank with NFTs, or the gaming industry trying to shift as fast as possible to profit from this new way of experiencing life.

A Metaverse can be portrayed as a new world where people interact with one another, regardless of how physically far apart they are. There are already manifestations of this phenomenon.

Games like Roblox, where independent developers can make popular games for children, might be the closest thing we have to a Metaverse. The fact that major events are held there, like concerts, art auctions, and even smaller events like business meetings (the Roblox team often has their meetings on the platform), gives an idea of what is yet to come. Fortnite is another game that almost accidentally stumbled on the

Metaverse or an early version of it. Initially, the game was just to be an online shooting game; then, one of its modes, Battle Royal, where players combat in teams while the map shrinks, became extremely popular, and Epic Games, the developer behind the game, was quick to turn the Battle Royale mode a lot more social by incorporating chats and other interactions. In fact, it is so social that a concert by hip-hop artist Travis Scott was viewed live from the platform by more than 12 million people.

NFT Technology

As a result, existing non-fungible token applications are only scratching the surface of the innovation.

Holders of non-fungible token technology get full ownership of digital assets. You will begin to comprehend how NFTs may create a Metaverse that connects society worldwide once you realize the security that possessing a digital asset on the blockchain provides. As shown by the amount paid for many digital art pieces, the notion of NFTs giving ownership over digital assets has already been embraced by the masses. The Metaverse is just the virtualization of that idea.

Some NFT initiatives are already going more and more into the Metaverse world. Most people consider blockchain gaming to be the model for an all-encompassing Metaverse. Most of the principles in the Metaverse have already been tested and modified in blockchain gaming.

Axie Infinity is the largest game project in the area right now and one of the largest non-fungible token projects generally. Players adopt, train, and battle in Axies, which are comparable to Pokémon. On-chain, each Axie is denoted by an NFT. As the Axies gain experience, they grow stronger and, as a result, costlier. It is possible to buy Axies and in-game products that may improve Axies' skills. In the realm of Axie Infinity, players will combat and interact with one another. Basically, Axie Infinity is a virtual environment where individuals worldwide communicate through digital objects. In some ways, it is its own constrained form of the Metaverse.

Another product that dabbles in the Metaverse is Xplorer's Studio, a collection of 10,000 non-fungible token astronauts. An Xplorer NFT can be used as a digital ticket to get access to a set of online tools that will assist in the creation of an online identity. Zoom backdrops, an exclusive store, a digital community wallet, and other unique features are among the functions available. In addition, and maybe more importantly, when the NFT transactions are completed, Xplorers Studio will acquire land in a Metaverse.

The purchase of virtual land in most NFTs projects is a major step forward for the Metaverse paradigm. It demonstrates that value can be derived from online real estate and adds to blockchain technology another aspect of our offline lives. This dedication to broadening horizons is exactly what Xplorer Studio attempts to instill in their community and NFTs.

The Metaverse is still in development. It is only now that the technology on which it will be constructed is being developed.

How to Get NFTs in Metaverse?

Users can buy Trending NFTs from legitimate NFT marketplace like OpenSea or Rarible. Users can also access these NFTs in their winning game markets. To buy NFT, the user must do the following:

Choose the hub where you want to buy NFT Metaverse.

When buying NFTs with ETH, make sure you have the correct ETH balance in your wallet. You can also buy Metaverse NFT through Metaverse Stones like AXS or SAND.

Pick a plug, browse the NFT Metaverse collection and pick the one that's right for you.

Tell the market about your wallet. Some NFTs require users to bid. Submit your offer and wait for confirmation.

If the offer is received, you will receive a receipt with the calculated amount indicated on the receipt created online.

6

METAVERSE INVESTMENT

Metaverse Business Application

The Internet gave rise to many new services we didn't have before, and these services have accumulated to provide us with what we call the second Internet era or Web 2.0. These services include social media networks, e-commerce, music streaming services such as *Spotify* and *Pandora*, crowdfunding platforms, live-streaming, etc.

However, with virtual reality gradually on the rise and more VR headsets being pre-ordered by consumers than ever before, it's no surprise that entrepreneurs are looking to jump on a new emerging market. According to a Zion Market Research report, the global augmented and virtual reality market is expected to grow to approximately USD 814.7 billion by 2025, far above its value in 2018 of 26.7 billion before the COVID-19 outbreak.

The Metaverse is still in its infancy but for those who act quickly and are patient to wait for the industry to mature, there any many lucrative business opportunities awaiting. Sooner or later, almost everything that's possible in real life is likely to be replicated in the Metaverse, in addition to new services and business models built on web3 technologies.

No one can predict what's going to happen in the Metaverse once it turns mainstream. For now, we can only speculate on what new business models are likely to emerge from here. Each person has their own prediction, based on previous experiences with similar technologies, but, ultimately, we do know it's going to be huge. So let's look at some of those business models already available and under development today.

Virtual Goods

The virtual goods trade is one of the more established business opportunities already thriving in the evolving Metaverse.

Virtual goods include digital clothing, shoes, accessories, weapons, furniture, appliances, and other objects that only exist in the virtual world. Virtual face masks, for example, were sold as wearables in *Decentraland* back in 2020 as part of a partnership with *Binance* to raise funds for fighting the pandemic. Second Life, meanwhile, has featured well over 2 million virtual items on its online marketplace.

Brand activation for well-known luxury and consumer brands is already underway, with Gucci opening a store in Roblox called Gucci Garden. One digital Gucci bag sold, for example, sold for $4,000 on Roblox in 2021. Other brands, including Adidas, are also following suit with expansion into the Metaverse and virtual goods.

In the future, brands venturing into digital fashion are expected to become more innovative and intelligent in designing items for wear in the Metaverse. Italian luxury fashion label Off-White's founder, Virgil Abloh, says he wants "to make virtual clothes to paint pictures physical clothes cannot, and let buyers access a new dimension of their personal style—no matter who they are, where they live, and the virtual worlds they love."

Virtual Malls

While online shopping is usually an activity done alone, this is about to change with the spread of virtual stores and malls. Shoppers will be able to shop for virtual clothes, visit their favorite brands, and peruse virtual art alongside friends from around the globe, all in one location.

Compared to the physical world, shoppers will be able to visit dozens of storefronts in dramatically less time and with less friction. One hypothesis holds that consumers will act more aware and responsive during the shopping experience without the friction of queues, closing hours, renting a trolley, finding a bathroom, staff shortages, and many other obstacles encountered in the real world.

One of the early contenders in the virtual retail space is Redfox Labs (www.redfoxlabs.io), which hopes to drive the adoption of virtual retail shopping in Southeast Asia. Consisting of 120 shopfronts (each created as its own NFT), Redfox Labs is opening a mall called Virtual Space where participating brands can purchase or lease a shopfront from other participants, including the ability to rent storefronts by the hour or for a designated promotion period.

Digital Art

The Metaverse offers an ideal platform for digital artists to showcase art in a 3D space and reach interested and cashed-up buyers. Artists can open their own virtual gallery in all the major worlds, including *Decentraland* and *Cryptovoxels*, where avatars can buy, browse, and even talk directly to the artists via voice chat.

In addition, the Metaverse provides artists with a platform to build a community and digital host events. This is one of the reasons why the popular Bored Ape Yacht Club NFT art project is making investments to acquire land and host events inside decentralized platforms. This includes building their own club on land purchased in The Sandbox, with admission reserved for those holding a Bored Ape NFT in their *MetaMask* wallet, similar to the exclusive Black Sun club depicted in the novel Snow Crash.

The world's oldest auction house, Sotheby's, has also built a digital replica of their *London New Bond Street Galleries in Decentraland's Voltaire Art District.*

Commenting on its opening in 2021, Michael Bouhanna, the Specialist and Head of Sales at Sotheby's, said, "We see spaces like Decentraland as the next frontier for digital art where artists, collectors, and viewers alike can engage with one another from anywhere in the world and showcase art that is fundamentally scarce and unique, but accessible to anyone for viewing."

Online-to-Offline and E-Commerce

In the web3 era of the Internet, many of the e-commerce websites we know and love today are likely to have a virtual presence in the 3D world—replacing part of today's reliance on 2D images and online catalogs of products sold on e-commerce websites.

Businesses, including food outlets, will increasingly look to adopt an online-to-offline (O2O) model by creating a presence in the Metaverse. *Domino's Pizza*, for example, has opened a small kiosk outlet in Decentraland where users located in the U.S. can order pizzas at their kiosk and then enjoy food delivered directly to their physical address.

At the same time, brands can leverage the many benefits of traditional online commerce by running an unmanned storefront 24 hours a day without traditional cost overheads such as coat hangers and mannequins. Dresses and other items can simply levitate in Metaverse outlets, unconfined by gravity.

Customers, meanwhile, can try on clothes and see items inside an immersive 3D rendered environment as part of a new buyer experience that starts in the Metaverse and connects with the real world. After trying on and paying for products in the Metaverse, consumers can receive their purchased items via the mail in the real world.

The ability to try on clothes in a virtual setting is also expected to reduce returned purchases (due to poor size fitting) and reduce the negative effect that returned items cause on the environment (i.e., wasted and non-recycled packaging, transport-related carbon emissions, and the destroying of returned items purchased via e-commerce). In the U.S. alone, up to 30% of all online purchases are returned. Return shipping generates over 15 million metric tons of carbon dioxide emissions each year (equivalent to the emissions from 3 million cars).

How to Invest in Metaverse

If you are a seasoned investor, you already know that it is never too late to get in on the ground floor. While most of what the Metaverse is might still be a mystery, there are currently ways for individuals to passively invest in Metaverse's prospects.

According to Bloomberg Intelligence, the Metaverse industry would be worth USD 800 billion by 2024, indicating that this may be a profitable field to invest in. We will look at four different ways you can incorporate the Metaverse into your investment plan.

Stocks in the Metaverse

Purchasing "Metaverse stocks" is one method that people interested in the Metaverse might invest in it. Holdings in publicly listed firms participating in the creation of the Metaverse are known as Metaverse stocks.

Considering that it is at the vanguard of developing the Metaverse, *Facebook* is an excellent example of a firm whose stock you may buy if you want to invest in it. Facebook's

CEO, Mark Zuckerberg, has stated that he does not want the firm to be identified as a social media firm. He likes it to be recognized as a Metaverse firm instead.

Suppose you do not wish to invest in Facebook. The social media firm already has an Oculus virtual reality headset and only recently introduced its first smart glasses in collaboration with Ray-Ban augmented reality to expand its AR platform. In that case, other businesses like *Microsoft, Unity Software, Roblox, Amazon, Walt Disney,* and *Nvidia* aim to play a part in Metaverse's evolution.

Metaverse Exchange-Traded Funds (ETFs)

Alternatively, investors may purchase a Metaverse ETF to acquire exposure to the rapidly developing Metaverse business. An exchange-traded fund (ETF) is a collection of assets that trade on a stock market.

Individuals can buy into a Metaverse stock ETF to invest in firms already creating the Metaverse or firms in a strong position to do so in the future. For instance, the Roundhill Ball Metaverse ETF (META) was developed to enable anybody to invest in and profit from the Metaverse. Investors may gain exposure to businesses including *Nvidia, Microsoft, Roblox, Tencent, Unity,* and *Amazon* with *Roundhill Investment's ETF.*

Virtual World Tokens

Virtual world tokens (VWTs) are virtual tokens that are associated with the VR landscape. Virtual world tokens may be used to purchase land and in-game treasures such as avatars in the virtual environment.

The Metaverse Index (MVI) token, for instance, is a virtual world token that gives owners access to a variety of tokens from cryptocurrency projects in fields including NFTs, virtual environments, and online gaming. The Metaverse Index token may be considered a Metaverse ETF for the cryptocurrency market.

Decentraland's coins are yet another example. Decentraland is a VR platform that enables users to purchase land, communicate with one another, and play games. It is the non-fungible token ecosystem's largest virtual environment, and its virtual land is reflected on the Ethereum network by the NFT LAND.

MANA and LAND are the two tokens used in Decentraland. LAND is an NFT that symbolizes all plots of land on the network, whereas MANA is the network's native governance token utilized for transactions.

NFTs in the Metaverse

The growth and acceptance of the Metaverse will be aided by blockchain-supported Metaverses that will most likely use non-fungible tokens and cryptocurrency assets. This is because a working Metaverse will enable people to navigate their avatars and virtual assets from one reality to another in a fluid and fast way.

Non-fungible tokens are digital assets that reflect a variety of unique products like art, in-game content, and collectibles. Individuals may now buy virtual pieces of land and even create their settings using non-fungible tokens, thanks to sites like Decentraland and The Sandbox. We should expect additional investment possibilities in this sector as we get closer to the Metaverse being a reality.

Investment Strategies

People were introduced to various types of entries as Metaverses normalized as a system that runs over the blockchain. As a result, investors can invest in both active and passive ways in the virtual world, depending on how it runs.

The user was actively involved in playing the game and participating in the virtual environment. As a full world, Metaverses had a plethora of categories, each with its own set of applications. A participant who participates in the game and is a part of the Metaverse earns money and NFT tokens (that the Metaverse runs on). These gained tokens are valuable in the financial world and can be traded on any Metaverse's marketplaces. Furthermore, users have the option of exchanging tokens for major cryptocurrencies.

Users can invest passively in the Metaverse while focusing on the active investing style. The NFT coin, which is utilized throughout the Metaverse, or NFT Metaverse, has a monetary value in the crypto world. As the project grows, it can list itself on various exchanges and platforms. Furthermore, investors can pool their money across the exchange or platform to gain profits because it is a part of several IDOs and launchpads. This brings us to the end of investors' passive participation in the Metaverse.

Another strategy for investing in Metaverses that might be recognized is investors' exposure by purchasing a Metaverse ETF. An exchange-traded fund (ETF) is a collection of securities and safeties that trade like stocks on a stock market. Investors can invest in companies with a strong position in the crypto ecosystem through a Metaverse stock ETF.

Passive and Active Investment Strategies

The passive technique is purchasing and storing coins rather than exchanging them regularly to avoid greater transaction costs. Because they believe they will not outperform the market due to its volatility, they prefer passive tactics, which are less hazardous. On the other hand, active tactics entail frequent purchases and sales of products. Investors feel they can outperform the market and earn higher returns than ordinary investors believe.

Value Investing

The value investing technique includes investing in a token based on intrinsic value rather than the market value since markets undervalue such companies.

Investors in such companies hope that, when the market undergoes a correction, the value of such undervalued companies will be corrected, causing their prices to skyrocket, providing them with substantial profits when they sell their shares. Warren Buffet, the world-famous investor, employs this method.

Income Investing

Coins in this sort of approach are chosen for their ability to provide cash flow rather than for their ability to grow the overall worth of your portfolio. Cash income from this form of investing can come in two types: fixed income (through yield farming) or dividend income (staking).

This technique is preferred by investors searching for a consistent income stream from their investments.

Investing in Dividend Growth Companies

Tokens with a track record of routinely paying interests are more stable and less volatile than other companies, and they strive to improve their dividend payout each year. In this strategy, the investors reinvest the earnings and reap the benefits of compounding over the long run.

Investing Guidelines for Beginners

The following are a few investing tips for beginners that you should consider before making a financial investment.

Set Financial Objectives

Establish financial objectives for how much money you will require in the upcoming period. This will enable you to determine if you need to invest in long-term or short-term investments and the amount of return you may expect to receive.

This will enable you to determine if you need to invest in long-term or short-term investments and the amount of return you may expect to receive.

Investigation and Trend Analysis

Before investing, take the time to thoroughly research how the market operates and how various financial instruments function.

Additionally, analyze and monitor the price and return trends of the coins you intend to invest in.

Portfolio Optimization

Choose the most appropriate portfolio from the portfolios that best fit your objectives. An ideal portfolio generates the most significant return while posing the least amount of risk.

Risk Tolerance

Determine the level of risk you are willing to accept to achieve the desired return. This is dependent on your short and long-term objectives as well. You would seek a more significant rate of return in a shorter period, while you would seek a higher risk in reverse.

Risk diversification is essential. Invest in different projects to diversify your risk and increase your returns. In addition, be sure that both tokens are not associated with one another.

Metaverse EFT

Metaverse ETF is a basket of assets that trade on a stock market. This ETF allows investors to participate in some organizations that are currently making the Metaverse a reality or poised to do so in the nearest future.

Matthew Ball, a futurist and trader, and other entrepreneurs, including Jacob Navok, CEO of Genvid Technologies, founded the Metaverse ETF. They claim to have done the legwork of studying the Metaverse, so you do not have to, and they are promoting the idea that the Metaverse can democratize everything in the process. In this scenario, they are decentralizing Metaverse investments by inventing an ETF that anyone can invest in and profit from Metaverse's rewards.

Bloomberg recently projected the market size of the Metaverse to reach $800 billion. With this ETF, investors would understand the worth of the Metaverse daily, even while it is still in its early stages. While the Metaverse is science fiction, it is conceivable to anticipate which firms will play a crucial part in bringing it to fruition.

An ETF assembles a bundle based on an index that includes a wide range, or as many as you like, of various public business equities. An investor can outsource their choice to this index if they want access to a subject or area but do not have the competence to choose all of the individual stocks."

Best Metaverse Investment Stock in Categories

Considering the technological challenges and financial benefits, it is doubtful that a single business would develop the Metaverse. The Metaverse must be so interactive that it is desirable to interact in a certain location or for a specified purpose. This will necessitate the use of technology from many sources.

High-performance computing, data services, next-generation networking, virtual networks, cryptocurrencies, and identity services, among numerous other technologies, will be required for a real version of the Metaverse.

Let's now take a look at the major technology categories for investment in the Metaverse.

Large Tech, Computing, Gaming, and Digital Assets are the four categories in classifying the investable assets associated with these sectors.

Big Tech Investment

- *Facebook* (FB)
- *Microsoft* (MSFT)
- *Alphabet* (GOOG)
- *Amazon* (AMZN)
- *Apple* (AAPL)

The FAAMG stocks, commonly known as Big Tech, are expected to handle the heavy lifting. They have the skills and resources needed to design Metaverse's architecture. Their key income sources (platforms, search, software, and cloud computing) will fund their Metaverse activities and function as critical Metaverse elements.

They also have the advantage of long time horizons, allowing them to 'fail' in ways that would obliterate smaller businesses. They are doomed to fail until they succeed. The 'disruptor' class of enterprises will invent Metaverse's features, while Big Tech will provide the framework.

Facebook (FB)

In June 2021, Mark Zuckerberg disclosed the next phase of Facebook, stating that "in the next couple of years," Facebook would shift from a social media firm to a Metaverse firm. On the other hand, Facebook has been laying the groundwork for the Metaverse for quite some time.

Thanks to Facebook's creation of a digital token for payments, the Metaverse will have an economy. What firm fully comprehends avatars better than Facebook? On Instagram, many users have already made avatars of themselves.

Facebook is in charge of bringing the real and digital worlds together. Facebook Inc. may be getting a new coat of paint. The firm is looking to change its name in the coming weeks, according to a report on technology news site the Verge, to indicate its recent focus on developing the Metaverse. According to nation world, the firm plans to hire 10,000 workers from the European Union to build the Metaverse over the next five years.

Microsoft (MSFT)

Metaverse applications are possible because of Microsoft's Azure cloud computing system. The firm's Metaverse technological stack is built on cloud and edge computing. Sam George, a Microsoft executive, recently wrote that they are best positioned to put this complete stack together. Their audacious objective is to make this stack more linked and smooth over time so you can simply travel up and down the layers that allow Metaverse apps.

George refers to Azure Digital Twin, a layer in Microsoft's Metaverse stack that can simulate an item, technology, or full area and keep the digital twins current and up-to-date using Azure IoT.

Assume that a Ford executive is attempting to identify possible obstacles at a factory in Detroit. Rather than boarding a plane and going to Detroit, he enters the Metaverse and finds himself on the factory's bottom floor. Real-time digital twins of machinery are in front of him, with analyses of their energy cost per hour, maintenance expenses, and productivity displayed above them. The CEO could make certain choices since he knows the digital representations are correct, courtesy of Azure Digital Twin.

The integrity of the Metaverse requires synchronization between the virtual and actual worlds. For clarity, Azure Digital Twins is simply one component of the firm's Metaverse technological stack. Microsoft is one of the few fully equipped firms for the Metaverse.

Google (GOOG)

Google is an intriguing Metaverse investment because of its exposure to the world's data and AI potential. Designing for the Metaverse is an excellent example of why Google restructured in 2015. Creating the parent company, *'Alphabet,'* enabled the firm to develop new technologies without detracting from its core businesses of search and marketing.

Google initiated a not-so-secret effort dubbed *Google X*, now known as '*X Development*,' in 2010 to improve life and commodities by a factor of ten. The *Moonshot Factory* is effectively a company within Alphabet's walls. X has spawned hundreds of various moonshot initiatives since its inception.

Google Glass, a VR-enabled set of goggles that eventually 'failed,' was one of these moonshots. Nevertheless, it was more of a public relations blunder than a failure. Google's interest in virtual reality and augmented reality is only the tip of the iceberg.

The firm is undoubtedly the most advanced in terms of autonomous driving technology. *DeepMind*, a *Google* affiliate, recently developed an AI technology to predict a protein's 3D structure based on its amino acid sequence. Google opted to make the information available to the scientific world to speed up research.

But do not forget about Google's main sources of income: Search, *YouTube*, and *Cloud*. Data, entertainment, and connection will define the next phase of the internet. For years, "Google" and "internet" have been interchangeable terms.

Given Google's vision, AI skills, and on-the-ground presence, it is difficult to conceive a scenario where the Metaverse exists without Google contributing substantially.

Amazon (AMZN)

Without computational elements and cloud technologies, a fully realized Metaverse is impossible. This is where Amazon enters the picture. Over a third of the public cloud industry is controlled by Amazon Web Services. It offers cloud computing services to the majority of businesses.

Firms who wish to have virtual real estate in the Metaverse would have to compensate Amazon for it, much like they did in Web 2.0. The gaming sector is a good illustration of this reliance. AWS is used by more than 90% of the world's largest public game firms.

The intersection of the Metaverse, gaming, and AWS will be a money-printing bonanza for Amazon. Nevertheless, there will be more possibilities in a decentralized world, and we assume Ether or an Ether-like blockchain to offer some type of structure in collaboration with Amazon.

Amazon and Google, we assume, will also contribute substantially to the Metaverse. It may not have a clear Metaverse plan like Facebook and Microsoft, but it will play an important role.

Amazon's current offerings already include physical and virtual elements. As the rate of blending increases, Amazon will reap the benefits.

Apple (AAPL)

To complete the finest FAAMG investment stocks, we have included Apple, but it is uncertain what function the firm will play in the Metaverse. In a decentralized Web3 ecosystem, the App Store's dominance over applications is threatened.

John Riccitiello, CEO of Unity Software, recently stated that the Metaverse simply could not operate as a gated community in a world populated with artists, designers, and innovators. The idea that any firm could pursue that goal is full of hubris.

Apple is also lagging behind Facebook and Microsoft in the augmented reality and virtual reality device industry. This put them farther behind in developing a Metaverse digital platform.

But you would be foolish to gamble against Apple; it has consistently delivered when we have questioned, "What is next?" The *Apple* world extends well beyond iPhone and Mac suites, with the iPad, AirPods, and the AppleWatch.

Hardware components will be required in Metaverse's initial phases, and we will not be left behind if Apple scores another home run in this sector.

Investment Stocks to Buy in Computing

- *Nvidia* (NVDA)

- *Advanced Micro Devices* (AMD)

- *Qualcomm* (QCOM)

- *TSMC is a Tai* (TSM)

Since Big Tech is the foundation of the Metaverse, the computer and semiconductor investments in this section are the foundation of the Metaverse's architecture. The most refined iteration of the Metaverse, as earlier said, is an entirely interactive one. Next-generation graphics will be required.

Nvidia (NVDA) and Advanced Micro Devices (AMD) offer CPUs and GPUs that allow high-level computing and expert visualization. Nvidia is, without a doubt, the greatest 'pure-play' Metaverse investment available. The AI and supercomputing required to make this a reality are powered by its main business products.

If you like to get even more involved, think about buying TSMC (TSM). This firm manufactures the chips sold by Nvidia and AMD. The last company on this list, Qualcomm (QCOM), is unique in this group.

Qualcomm develops wireless-related chips, software, and services. Its processors deliver high-performance processing for mobile apps and devices like smartphones, robots, virtual reality, and augmented reality.

Best Gaming Investment Stock

- *Unity Software* (U)
- *Roblox* (RBLX)
- *Sea* (SE)
- *Zynga* (ZNGA)

The game industry has a greater understanding of the Metaverse than anybody else. These firms are the gateway into the Metaverse world; they have been developing virtual environments for years and have made numerous variations on what works and what doesn't.

This tendency has been driven by the rise of in-game engine companies such as Roblox (RBLX), Unity Software (U), and

Epic Games, which empower players. Participants may utilize these systems to play video games and develop them using no-code applications.

The user has progressed from enjoying the game to producing it and will ultimately be a part of it. Innovation is decreasing the bandwidth. We will scoff at the thought of using a controller to enjoy a game in the future.

Our understanding of the term "video game" will most likely evolve as well. We will stop thinking of them as "games" and start thinking of them as "experiences" as bandwidth reduces and gets more interactive.

Digital Assets Investment Stocks

Square (SQ)

PayPal (PYPL)

Bitcoin

Ethereum

The Metaverse and cryptocurrency were built for one other. We understand that money is a technology that will alter dramatically as Web3 transitions. There is potential there. The Ethereum Blockchain technology has the potential to enable initiatives like DeFi apps that eliminate the need for banks and other financial intermediaries. Non-fungible tokens might be used to establish digital asset ownership.

Bitcoin's infrastructure and monetary policy might flourish in a more digital world, reducing the Federal Reserve's role. Layer 2 technology, like the Lightning Network, will allow for quick and safe fund transfers between two entities without requiring third-party control.

The price of Bitcoin and Ethereum would benefit from these uses. Other people will desire to utilize those valuable products as more developers build on their networks and generate great offerings. They will require the blockchain governance token to pay for those offers. Thus they will purchase bitcoin as an asset.

Because bitcoin's demand exceeds supply, continuous or growing demand for the asset should drive its price to rise. Square (SQ) was also added to this category since it is a publicly listed Fintech business that fully embraces the possibilities of cryptocurrency.

In the future years, PayPal (PYPL) will do everything Square does. It has the potential audience and resources to steal Square's technology without paying a cost, i.e., it may be second to new products and services.

The Cash App and PayPal's Venmo will almost certainly support it in the Metaverse, irrespective of traded money.

Other Investment Stocks to Consider in the Metaverse

- *Disney* (DIS)
- *Snapchat* (SNAP)
- *Zoom* (ZM)
- *Twitter* (TWTR)
- *CrowdStrike* (CRWD)
- *Okta* (OKTA)

The majority of technologically successful businesses will gain from the future age of the internet. Thus our honorable mention list is in no way complete. The diversity of the list, though, reveals the Metaverse's second and third-order benefits.

In virtual environments, Disney (DIS) can digitize its parks and release the world's most powerful IP portfolio. Snapchat's (SNAP) augmented reality, virtual reality, and critical messaging services will seamlessly move into the Metaverse.

Do we honestly believe Zoom (ZM) will stop advancing in the field of 2D video conferencing?

Twitter (TWTR) is experimenting with small transactions for content in a Web3-like approach. As a growing amount of your assets sits in the virtual vs. physical realm, cybersecurity stocks such as CrowdStrike (CRWD) and Okta (OKTA) will become significantly more relevant.

Although the Metaverse has the potential to cause a stir among the FAAMG firms, the most powerful technology businesses today will lead and benefit from this shift, albeit to differing degrees.

Metaverse Investment Stocks Alternatives

Purchasing a bundle of Metaverse investment stocks in an ETF is an alternative to purchasing specific Metaverse stocks. Like Invesco QQQ or ARKK, any tech-heavy ETF will offer you enough access to these names.

The Roundhill Ball Metaverse ETF (META). This is among Roundhill's six ETFs that concentrate on future investing themes, including eSports, streaming, sports betting, and various other topics.

As you can see, there are several methods to become acquainted with the Metaverse.

Marketing in the Metaverse

Digital marketers must keep up with the most recent technology advancements. Knowing the Metaverse and its potential is part of this. Marketers have to realize that the Metaverse is not simply a fad; it appears to be here to stay and on its way to becoming the next big thing.

What strategies may marketers use to adapt as the Metaverse grows? First and foremost, marketers must remember the importance of millennials and Generation Z as a target market. Certain kinds of Metaverses, like games such as Roblox and technologies like virtual reality, are also popular among these generations. Let us look at how advertising can be conducted in the Metaverse.

Parallel Metaverse Marketing within Real-Life Marketing

Develop promotional experiences that connect with real-life events or are similar to what your company presently does in the physical world. For instance, AB InBev's beer firm Stella Artois worked with Zed Run to develop a Tamagotchi-inspired experience coupled with the Kentucky Derby. They did so since Stella Artois, a brand of AB InBev, is known for supporting sporting events, particularly horse racing. As a result, developing a virtual platform where NFT horses may be sold, raced, and bred appears to be a natural next step for them.

Immersive Experience

You can provide digital marketing in the Metaverse space. Bidstack, a game ad tech firm, shifted from physical-world outdoor marketing to virtual billboard marketing.

However, virtual billboards are not the only option. Instead of merely posting commercials, offer branded installations and events that people may engage with. Because Metaverses are engaging and interactive by nature, it is ideal to capitalize on this by providing a similar interactive experience with your promotions and marketing efforts.

We have seen early adopters provide an interactive experience to their consumers, like a Lil Nas X performance in Roblox, Gucci Garden experience visits, and Warner Bros.' advertising of 'In the Heights' with a digital version of the Washington Heights neighborhood. Partnerships with the Roblox Metaverse and other Metaverses have now shown new revenue sources for businesses.

Make Collectibles Available

Individuals like collectibles, and the Metaverse provides them with another platform to do so. You may reproduce the experience in the Metaverse by offering products or unique items that can only be obtained in the Metaverse.

The Collector's Room, for instance, is available in the *Gucci Garden Roblox* experience. In the Metaverse, it enables users to gather limited-edition Gucci products. Gucci made a total of 286,000,000 Robux from the game's early sales of collectible products.

Engage with Existing Communities

The public generally dislikes advertising. As businesses aim to break into the Metaverse, they mustn't irritate those who are already there. You will also need the favorable feedback of these users because you will be marketing to them.

Note that you may not simply enter a new system without considering the new layout. For instance, in Roblox, businesses gain more traction when they collaborate with members of the Roblox developer community to create products and experiences. Likewise, when O2 put up a performance on Fortnite, they teamed up with developers who were experts on the game.

Consider this a form of influencer marketing. Community members become key aspects of the implementation of the marketing since user-generated content is vital.

Continuously Experiment

Marketers are in an incredible time. While some core principles can help marketers determine what techniques and methods to use, the Metaverse is still a relatively young platform with plenty of potential for experimentation. Best practices are still being defined, and paradigms are still being developed in their entirety. This provides marketers a lot of flexibility to explore new things and be creative in their strategies.

Other Ways to Get Paid in Metaverse Creator Economy

Virtual Avatars

In many virtual environments, having an active avatar representation provides users with a sense of self. It is a natural human desire to express oneself through personal style, including hairstyles, colors, make-up, piercings, tattoos, and other physical features to clothing and accessories. In a virtual environment, the human body and any accessories have significantly more creative potential, and the creation of avatars, skins, and accessories is a rapidly expanding field. Roblox is an example of a world featuring a person-to-person avatar and accessory marketplace.

Apart from avatars in virtual environments, there is a contemporary trend of generating and selling profile images for use on Twitter or other digital platforms. Human images are popular right now, but so are images of other animals. The most common is the generative art style, which involves creating a set of thousands of versions of unique combination rarities by randomly mixing a set of many different body shapes and accessories/modifications. Cryptopunks and Bored Apes are two well-known examples of PFP projects that have jointly sold for huge amounts of money.

CryptoPunks have become a form of a cultural phenomenon that the cheapest punk costs 102 ETH ($391k). CryptoPunks' founder, Larva Labs, has secured an agreement with UTA to represent them in content agreements for films, games, and other projects. Punks are on the verge of evolving into ideal virtual beings.

Avatars come in different styles, from pixel art to cartoon-style 2D to powerful 3D representation with animation abilities. The proper style will depend on where you intend to sell the avatar for use. You have a leeway to show your creativity and unique expertise through these numerous channels.

Using non-fungible tokens to package avatars simplifies the process of establishing and selling ownership rights. Games that are not built on the blockchain network have their marketplaces and standards.

Virtual Products

There are numerous options for what a virtual product might be and what kinds of value it can provide to entice people to purchase it. A single item could fall under more than one category.

People buy virtual products for different reasons, including creative art pieces, collectibles, investment, memorializing and proving involvement in an event, expressing support for a movement, or just looking cool.

As with the CryptoPunks, the social standing of being an informed early adopter or a heavy purchase cost supporter of a unique release collection has become a new source of worth.

Goods with Game Utility

Aside from just aesthetic or social value, many emerging experiences go even further in terms of utility. The Sandbox game goes beyond what Roblox or Neon District can provide by enabling game developers to create creatures with statistics and other creative elements to inhabit their space.

Virtual goods for blockchain-supported environments can likewise be packaged and sold as non-fungible tokens, although they usually come with additional world-specific limits. Interoperability of virtual products between environments is a hot topic in theory, but it takes a different approach in practice. It is best to determine which virtual space you want to make for first and then look into their method.

Creators Tools/ Creator Training

Many potential gold miners incurred losses during the Gold Rush, but those who sold picks, shovels, tents, and blue jeans (Levi Strauss) earned a decent profit.

While everybody is racing to generate content in the hopes of a hit, if you have both innovative and technical talents, you can develop tools, set up training classes, or build communities to help many individuals build more efficiently. 2D/3D animation technologies are typically sophisticated, requiring substantial experience and expertise, costly, or both. It is not easy to set up a model for animation with natural articulating joints and control points. Creating more simple and purpose-driven tooling or assisting individuals in learning how to get the most out of current tools with exactly what they have to know has a business opportunity.

Content Remastering

Many 2D avatars and profile image projects are competing for attention in an increasingly congested market. Most are opting to provide 3D-built models for usage in one or more virtual environments or games to offer their owners greater value and the chance to connect as a community.

CyberKongz has been remade as 3D voxel characters for use in the Sandbox. Digital artists are in high demand to produce classic or voxel-style 3D copies of the 2D character pictures and supporting animations. Approaching non-fungible token projects that are either going to debut with eventual 3D editions or have already debuted with the promise of a later release of 3D editions are fantastic spots to engage and use your imagination and expertise to help them advance their early concepts.

Concerts/Live Social Events in Virtual Environments

Millions of individuals may be found online in many of the most prominent games and virtual environments. Famous musical acts have made tens of millions of dollars performing concerts in these environments.

According to Paul Tassi of Forbes, merchandise sales from Fortnite with Travis Scott totaled $20 million for the artist. That is compared to his $1.7 million one-night records for an in-person Astroworld tour and little under 40% of what he recorded earning from the $53,5 million tours.

Artists with limited fan bases can still host events in virtual environments to promote digital record sales, improved streaming, and commemorative digital merchandise. Yes, the decades-old concept of a concert T-shirt could now be realized as a non-fungible token or a game skin.

Although not directly creating and selling a digital product, roles associated with developing a digital event experience will be highly demanded. Some examples are listed below:

- Director
- Broadcast Producer
- Sound Engineer
- Artist in 3D

- VFX Artist

- Lighting Artist

- Motion Capture Actor

- Motion Graphics Designer

Metaverse Crypto

The Sandbox (SAND)

This is an NFT-based Metaverse where users can play, build, earn and monetize their assets in the virtual space. This virtual platform built by the Sandbox is community-driven and a significant representation of the basic principles of decentralization. The user-generated content environment of the platform has its own NFT marketplace where users can buy and sell in-game land and estate with a more immersive experience. Furthermore, the transaction inside the platform is carried out through its official token, $SAND. Also, the $SAND token holders are given access to the platform's exclusive features alongside their governing rights and staking.

Decentraland (MANA)

You ought to know that MANA, a Decentraland's token, is another Metaverse token. Decentraland is a VR platform built on the blockchain network that allows participants to purchase, trade, and develop land while playing games, generating content and communicating with other participants.

The largest virtual world in the non-fungible token realm is Decentraland, denoted by the non-fungible ERC-71 token

LAND. On Decentraland, each piece of land is unique, and its owners have complete control over what they do with it.

MANA is a platform that allows a participant to buy land. MIt is likewise based on Ethereum, and it may be purchased and traded in exchange for other crypto or fiat money. The overall supply of MANA has been set at 2.6 billion.

Enjin Coin (ENJ)

You might also check out Enjin Coin, which is a Metaverse token. It is an Ethereum-based token designed to use non-fungible tokens as simple as possible for people, brands, and enterprises. Non-fungible tokens developed using Enjin, on the other hand, employ the ERC-1155 standard, which is not the same as the ERC-721 standard.

ENJ is the governance token on the platform. ENJ directly backs non-fungible tokens that are created in the Enjin ecosystem. A fixed quantity of ENJ is created into each new non-fungible token created on the network. The locked funds determine the freshly created tokens' physical-world value. Enjin also received about $19 million lately, which will be utilized to develop a Polkadot-based blockchain for non-fungible tokens. A total of 1 billion ENJ tokens are available.

Render (RNDR)

These computer-generated environments require graphics, which necessitate a significant amount of computing power. Render is a decentralized platform that generates pictures by utilizing idle graphics processing units (GPUs). Users in the platform gain tokens, and Render may supply businesses with cutting-edge graphics in an economical and scalable manner.

Wilder World (WILD)

This token is native to Wilder World WILD, a Metaverse created with the Unreal Engine 5 game engine on the Ethereum blockchain. Digital goods can be purchased while it is under development and the goal of being a non-fungible token marketplace characterized by high liquidity, decentralized and open to the community.

- It has great functionality and allows you to design and trade your own NFTs.

- It is currently trading at $3.88 and has a market value of $325 million.

7

METAVERSE AND REAL ESTATE

What is Metaverse Land?

Land in the Metaverse

Is there anything special about Virtual Land, and what can you do with it?

True, it might "just" be a virtual metropolis, but investors are forking over real money to buy land in it. Buyers in Decentraland have the freedom to build anything they wish on their parcels of land. To many, trading products and services for bitcoin is a way to generate money in the virtual world. The American dream is taking on a new meaning in virtual worlds, where investors purchase pieces of property and strong communities increase desirability.

The American dream is taking on a new meaning in virtual worlds, where investors purchase pieces of property and strong communities increase desirability.

Why do People buy Virtual Land? People buy virtual land for a variety of reasons.

The following are a few of the reasons why purchasing virtual land is deemed necessary:

New Asset Class

Digital real estate has proven to be a valid asset class. Its value is expanding exponentially, which makes it an appealing investment possibility. There is also a possibility that someone may turn it into a viable financial asset, analogous to real-world art and actual-world real estate.

Fear of Missing Out (FOMO) (Fear of Missing Out)

People who purchase wonderful reals or virtual do so because they have a horrible sense of missing out on something amazing. So many individuals miss out on buying Bitcoin when it was so cheap has prompted them to explore alternative items such as virtual real estate.

Exceptional Profits

Because of its association with the constantly increasing crypto-investment world, virtual land has the potential to generate tremendous returns. Many people have been able to gain thousands of dollars in a short amount of time because of the ease with which they can flip land (like they do with real estate) and the consistency of the bull market.

Possibilities for Additional Earnings

In the long run, virtual land opens new possibilities for what can be done with the property, such as constructing art galleries, executing advertising campaigns, or simply renting the land to others to earn money from their construction projects.

Some users utilize their virtual property to build virtual casinos, in which they can play.

In addition, major retailers are looking into the prospect of opening storefronts in virtual reality (VR).

It is Less Complicated than Purchasing Real Estate

There are also significant advantages to using digital real estate over conventional methods, including eliminating time-consuming paperwork, property maintenance, and tax payments. Furthermore, the usage of blockchain technology enhances the security and traceability of land acquisition transactions.

Low Entry Barrier

Prices for real estate are rising worldwide, yet virtual properties provide similar benefits for less than 1% of the cost. Due to the high cost of acquiring natural land, this goal may be out of reach for many people, but purchasing virtual land is lower.

Buying Virtual Land in the Metaverse

To a large extent, earnings outweigh all other considerations when it comes to purchasing virtual estate (as well as a cryptocurrency or NFT).

Numerous newcomers to this ecosystem are looking for other means of earning money in the wake of the coronavirus outbreak. Even though some people are simply trying to save up considerable sums of money obtained via cryptocurrencies like Bitcoin, Ethereum, and Solana, these factors certainly contribute to the current bull run for virtual land.

Although it may appear to be a pointless endeavor to some, people are spending millions of hours every day in virtual worlds despite this.

Thus, it will retain its value as long as others believe it is valuable. Some short-term investors will indeed be interested in profiting off virtual plots in the hopes that their value will increase in the future. Another group of people, on the other hand, will just seek access to these plots because they love doing so.

Advantages, Disadvantages, and Risks

Advantages of Buying Digital Real Estate

- Diversification simply refers to investing in a variety of assets. You can buy multiple websites and domain names in the context of digital real estate. Then, utilizing the ways we described before, produce passive money.

- **Profits in the millions:** If you're lucky, your website might be sold for millions of dollars! That is why some people are attempting to acquire as many domain names and websites as possible.

- **Self-employment:** This may help you become your employer! Buying and building a website might generate more money.

- **Scalability:** Did we mention that this might be turned into a business? Your only restriction is the number of individuals that have access to the internet!

Disadvantages of Buying Digital Assets

- **Volatility:** A variety of factors might jeopardize your digital real estate. Take, for example, the Google search engine. It can affect how websites appear on search results pages.

- **Required abilities:** If you don't understand how a website or domain name works, you'll have difficulty making money from them. Building a website and generating traffic for it requires talent. However, you might study from a variety of online programs. However, you will need time to learn.

A competent digital marketing agency will take care of the SEO optimizations and website creation for you for a charge! Before cryptocurrencies, investment in digital real estate was like that. People are already buying digital plots of land now that it has become trendy.

Decentraland is the most popular right now. The value of its real estate is measured in LAND tokens. There are two methods to invest in this form of digital real estate. The first is to create your land on the blockchain. However, specific creative abilities, such as 3D modeling, may be required. The second step is purchasing and storing the tokens.

Risks

Unlike investing in the real estate market, where your acquired physical land is guaranteed to survive, digital land in the Metaverse will become non-existent if the platform you purchased fails and goes down. Another thing to remember is the significant volatility of the cryptocurrency used to transact in the Metaverse's real estate market. Because the value of digital money varies, the value of the Metaverse property you own fluctuates accordingly.

Furthermore, because digital real estate is a comparatively new asset class, many facets have yet to be explored. As a result, investing in the Metaverse's digital real estate market is very speculative; thus, thoroughly researching the advantages and downsides is recommended before making any decisions.

8

METAVERSE AND MUSIC

Create and Perform in a VR

The Ability to Create and Perform in a Virtual Venue

This can be through the use of an avatar or other visual representation of the artist, sometimes mixed with an authentic video representation of the artist.

So, this requires new production capabilities, such as manipulating the virtual environment and combining digital visual production with the artist's musical production.

Interactive in Real-Time

The performance by Travis Scott on Fortnite was likely the most striking and commercially successful example of this revolutionary musical art form in recent years.

The performance by Travis Scott on Fortnite was likely the most striking and commercially successful example of this revolutionary musical art form in recent years.

This event generated a significant amount of attention and interest for this event.

Aside from virtual events and NFTs, another Metaverse trend that has impacted the music industry is the emergence of virtual "artists."

The thought of listening to a virtual artist, who is made by artificial intelligence and does not have a real personality, may be repulsive to many serious music enthusiasts.

Despite this, there is no disputing that such musicians are gaining significant traction among young people who grew up with the Internet. The rapper FN Meka, who has been described as a "robot rapper known for his flamboyant style and Hypebeast aesthetics," is an excellent illustration.

It is built on a foundation of serious commercial possibilities. While writing this book, the virtual rapper has over 9 million followers on the TikTok video-sharing app.

Comparatively, Chance the Rapper, who is sometimes referred to as "one of the new crops of superstar rappers," had less than 2 million TikTok followers when writing this book.

Is the Metaverse an Opportunity or a Danger for Music?

Both opportunities and threats for the music industry can arise from the Metaverse, as demonstrated by the two cases presented above. Artistic careers are at risk if they rely on outdated methods that are no longer relevant in today's world of cutting-edge production and consumption methods and consumer experiences. For example, suppose you only possess the rights and monetize through subscription streaming channels. In that case, you won't be making enough money to justify your investment in these methods. They'll quickly become commoditized and automated.

Business opportunities are virtually limitless for those willing to push the boundaries and use all available technology to interact and create. Compared to online Metaverse performances, even the most extensive arena tours cannot handle anything near the instantaneous, one-time global crowds the artist can attract to a live online Metaverse performance.

The COVID-19 pandemic, which caused the entire world to shift to the Internet for entertainment, has demonstrated to the music industry that ticketed, well-produced, and compelling live streaming will be around for the foreseeable future. It is conceivable that the most significant concerts and festivals that take place in the actual world will in the future have an online component that is more committed, sleek, and transactional. Because of this, the Metaverse will continue to exist in music for the foreseeable future.

Because of this, the Metaverse will continue to exist in music for the foreseeable future.

Legal Ramification of Music Being Played in the Metaverse

When music is generated, played, streamed, and exploited online, rights clearances are the most important consideration, as they are in all aspects of the music industry. Most of the standard legal and licensing regulations for online exploitation apply in the Metaverse, with some exceptions.

For example, a digital music service provider (such as *Spotify*) may promote and organize a live-streamed concert on a worldwide games console platform (such as the *Sony PlayStation*).

This concert could occur during the tournament's intermission being held and marketed by a leading games publisher (such as *Electronic Arts*) who might be collaborating with a well-known brand during the interval of the event (e.g., *Adidas*).

Those interested in attending would need to be registered users of the gaming platform and have acquired entrance tickets to the eSports competition. Although the live-streamed concert would be available to a restricted number of superfans who joined a prize drawing by purchasing an original NFT token issued by the headline performing artist, the performance would only be open to the general public (for example, Drake).

Top-level prizes may include attendance at the live virtual event and an actual piece of digital goods.

Runners-up would still be able to watch the concert on-demand later, even though they'd be missing out on the thrill of a live show. As illustrated by the example above, the network of contractual responsibilities to negotiate and the rights-clearance concerns are not unlike the issues that lawyers may encounter in the real world when dealing with clients. The half-time show for the NFL Super Bowl is well-known in the music industry for being a highly prestigious but demanding production and clearance exercise that requires much planning and coordination. However, in many ways, the amount of complexity associated with clearing music for the Metaverse can be substantially higher than the level of complexity related to clearing music for the physical world.

Therefore, anyone wishing to use another's music in the Metaverse must ensure that the terms under which they receive a license are compatible with where it is being utilized. Censorship and content standards impacting a live performance by a Top 10 rap artist in the United States will be drastically different from those affecting a similar performance in, for example, Indonesia, Dubai, or Hong Kong. The political beliefs of artists are frequently expressed onstage.

These situations are more manageable in real life. Still, they are the stuff of nightmares for the legal compliance teams at large platforms, frequently tasked with maintaining positive relationships with local governments worldwide.

9

REQUIREMENTS OF METAVERSE

Advantages of Business-Owned Metaverse

Metaverse development is the most creative and commercially viable option for future-oriented businesses. Let us have a look at why this is so:

Meta-Versatility

Being able to respond to changing needs and modify things as much as possible without consulting the system is a huge advantage for businesses. Do you need to keep up with the most current product lines? Simple. Do you want to establish any unique breakout areas? Make it happen.

Thanks to this flexibility, businesses may customize their content and environments in any way they want. They can build bizarre, otherworldly landscapes that can be continuously improved, altered, or even fully recreated in a short amount of time.

Businesses can seamlessly keep up with the speed of culture while maintaining complete creative control by becoming owners of their world.

Keep it On-Brand

Controlling content also implies that the digital environment as a whole can be kept on brand. Although Metaverse participants may have to meet the standard set by Roblox or Fortnite, Metaverse builders may define their limits and have complete control over how their business is viewed, from layout to visuals to activities and messages.

This enables virtual users to fully submerge themselves in the brand without the possibility of compromise from third-party platforms.

All Under One Virtual Umbrella

The most significant advantage is that owned Metaverses enable all business operations to be centralized in one location. Different applications for every new campaign, ephemeral microsites for unique items, and online events for different time zones are no longer necessary. Everything can exist together in a readily accessible area for participants worldwide to engage and enjoy 24/7 with a virtual brand world.

Because everything is enclosed, there is only one path to buy, making it easier than ever for Metaverse participants to meet, communicate, play, explore, and make purchases—all in one location.

Expand Your Tribe

A brand Metaverse could achieve the same result in the virtual realm as a purpose-driven physical flagship can do in the physical environment, with one crucial distinction: the reach will be far greater than in a physical world context.

This implies that companies may grow and build their global audience by giving people a permanent shared place to communicate with one another and the company while discussing their hobbies and interests. The power of a unified tribe behind a business cannot be overstated, and growing and nurturing these links would be a wise strategic option for any commercially focused business.

Potentials and Future Opportunities

Even though the Metaverse comes short of science fiction authors' fanciful dreams, it is expected to generate trillions of dollars in worth as a novel computer platform or content medium. However, in its complete form, the Metaverse serves as a portal to most digital experiences and an essential piece of all physical ones and the next major labor network.

The benefit of being a significant player, if not a driver, in such a network is self-evident. Still, virtually all of the biggest online businesses are among the world's top ten most profitable public companies. And, if the Metaverse truly serves as a viable "successor" to the web, with much more reach, time spent, and economic activity, there will certainly be even more economic benefits. Nevertheless, the Metaverse should provide the same breadth of potential as the web—new firms, products, and services will arise to handle everything from payment systems to identity verification, employment, ad distribution, content development, and security, among other things. As a result, several incumbents are expected to lose their seats.

The Metaverse, in general, has the potential to change how we distribute and commercialize modern resources. Industrialized economies have changed as labor, and real estate scarcity grew and fell for millennia. Would-be workers living outside of urban centers will engage in the "high value" economy through digital labor in the Metaverse. We'll see more adjustments in where we live, the constructed architecture, and who do particular activities as more consumer purchases switch to virtual products, services, and experiences. Consider the concept of "Gold Farming." Many "players"—often hired by a bigger firm and generally from lower-income nations—would devote a workday accumulating digital items for sale inside or outside the game not long after in-game trade economies began. In the West, these sales were generally to higher-income players. While most "labor" is tedious, repetitive, and restricted to a few uses, the variety and worth of this "work" will expand in tandem with the Metaverse.

Evolving of the Metaverse

Outsiders frequently mix up Metaverse and Virtual Reality. They believe Metaverse is a new form of virtual reality technology. Although it may appear to be science fiction, Metaverse digitally combines personal and business life in a way that is similar to our physical world. Some people believe it is the internet's future. Even if these notions are correct, the digital world is evolving. However, you may be wondering why this technology is capturing people's interest and why they are investing so heavily in the digital world.

Unlike virtual reality (VR) technology, which we utilize in video games, Metaverse incorporates all potential activities. You can do everything online, from hanging out with pals to going to the movies, playing tennis, and going to concerts. Take a closer look at how Metaverse will transform our world:

1. Economic Changes

Metaverse will alter the way we do business in the future. It has an impact on how people think when they buy things. As a result, companies will conduct more market research to understand their customers' behavior better. The experience of buying things in a physical store is not the same as what Metaverse provides. As a result, every company must upgrade its operations. Customer contacts will undoubtedly be handled by robots and virtual assistants. For data analysis, these bots will be equipped with powerful computing equipment.

2. Cultural Changes

Metaverse will affect cultural standards since it links people from many ethnic backgrounds. People in the Metaverse will have connections and friendships just like they do in the physical world. They do, however, engage through holograms and self-contained NPCs. The Metaverse will have an impact on the corporate world and will bring customers together in 3D. They won't be able to communicate with marketing people, but they will be able to communicate with bots to get answers to their questions.

3. Shopping Experience

In comparison to physical purchasing, the Metaverse shopping experience is unique. In the Metaverse, virtual real estate, avatar skins, and virtual fashion have great value. People will also invest in enterprises and properties that do not exist physically. Because people will use avatars to represent themselves, the fashion industry will focus on designing clothing for the characters. People would also look for virtual designer clothes and mansions to invest in the Entertainment Industry

In the Metaverse, virtual concerts, seminars, and gatherings will be commonplace. Celebrities and brands will use the virtual world to interact with their fans. We now use technology to make purchases and play games. On the other hand, people would virtually spend time with their pals at restaurants, events, and cafés. Wendy's, for example, is experimenting with putting their restaurant in the Metaverse so that consumers may engage with their friends there. Ariana Grande's concert in Fortnite Metaverse on August 7, 2021, is another example of a Metaverse entertainment event.

Challenge of the Metaverse

Till now, we don't have a clear idea of what the Metaverse space will look like, but we can expect it to face the following challenges to become a full functioning virtual environment:

Reputation and Identity in the Metaverse

When we speak of the physical world, the issue of personal identification and representation is quite easy. However, when it comes to virtual worlds or the Metaverse, one can question what exactly constitutes one's identity. And, most importantly, how to show that you are who you say you are, rather than someone or a robot attempting to simulate your identity. This is where reputation comes into play, not only in identification but also in verifying that the party with which one interacts is reputable and legitimate. The ability to manufacture facial characteristics, footage, and voice poses the biggest hurdle. Therefore, we can expect new identification methods to emerge in the coming years.

Data and Security in Metaverse

Although businesses and organizations continue to improve their internet security systems, data protection has long been an issue for individuals in many online spaces. Delving deep into the Metaverse will necessitate the evolution of security systems to a whole new level to keep up with this ever-expanding environment. This would require developing new personal data and data protection systems to ensure the security of one's identity and belongings in the virtual environment. As a result, personal identification may reach a stage where participants must supply more personal information than is now required to identify themselves and verify that the security system functions properly, keeping personal information safe.

Metaverse's Currency and Payment Systems

Bitcoin is one of the most popular examples of digital currency, which has been around for several years. The same can be said for online markets such as Amazon and eBay, which link millions of people worldwide. Without a doubt, Metaverse would have its online marketplace, integrating several physical and digital currencies for quick and easy trading. It will be particularly important to build a unique new transaction validation mechanism, irrespective of the currency or marketplace structure when it comes to transactions. The challenge will be persuading participants that they can trust and, more importantly, feel safe when trading in the Metaverse space.

Law and Jurisdiction in Metaverse

Identifying jurisdiction and laws that can guarantee the virtual environment is safe and protected for its users will be a major task. The subject of law and jurisdiction will arise due to countries' immersion in the Metaverse, necessitating a greater focus on virtual legal areas. With the rising virtual world available to people worldwide, it will be necessary to determine how the jurisdiction will be applied. The Metaverse is poised to attract a huge number of people together, making it a terrific platform to interact and exchange information. Still, it also puts participants at risk if no regulations govern the boundaries.

Ownership and Property in Metaverse

When we talk about a single integrated virtual environment where you can connect with the world and other people, just like in the physical world, we may envisage buying and selling different products and assets. NFTs have accelerated their rise to prominence, creating waves in 2021 and drawing more investors and consumers to digital assets and tokens. The task will be to create a uniform platform that can authenticate the owners of digital assets in the Metaverse, comparable to how non-fungible tokens already depict physical-world items, providing and verifying ownership rights for artwork, music, films, and much more.

Metaverse's Community and Network

Without question, Metaverse space will unite a varied group of people together, bringing individuals from all over the world together in a single virtual reality. As the past couple of years have demonstrated, being interconnected is necessary for people, forming a powerful network in the Metaverse for business and personal reasons. Metaverse can become a technology for several people to connect and build genuine relationships. We are used to talking via the internet, but for the Metaverse to become a place where users can feel mentally and physically present, haptic and motion capture innovation will need to progress to a whole different level. This level will allow for a great understanding of one's existence and environment and visual fidelity, and the capacity to touch and feel.

Time and Space in Metaverse

When comparing the physical world to a virtual world, the idea of time perception can be varied, as participants seem to be less aware of their bodies while within the VR. People may subconsciously devote considerable time within the Metaverse due to the full interaction. Given the risk of a skewed sense of time, it is critical to implement systems that keep participants in touch with reality.

The concept of space is another concept worth considering in the Metaverse world. Because the Metaverse implies an unlimited space, it may be difficult for participants to submerge themselves in such a massive world at first, trying to absorb the volume and myriad of information available all at once. To guarantee participants are both aware and secure while within the virtual world, both time and space interpretation in the Metaverse will need direction during the first phases of immersion.

And although we are still on the verge of creating a virtual world, setting up the Metaverse might be both difficult and rewarding. It is unquestionably necessary to ensure that the

Metaverse serves as a supplement to the physical world we live in rather than a replacement. As exciting as the virtual world is, it is critical to be informed, protected, and secure in this new huge environment.

Potential Legal Issues

There is a chance of practising "fraud" in digital activities. One who deliberately deceives someone to convince them to modify their position to their detriment is liable for any damage they receive.

However, the conduct of fraudulent transactions may be increasingly difficult with the arrival of these advanced technologies. On the other hand, criminals always devise a method to profit from technical flaws. In principle, the law lags behind technology, which will continue indefinitely. This is the case because technology progresses quickly, and the law usually follows suit. The legal system develops new rules by issuing court judgments at the administrative, state, and federal levels. The legislative body prepares and promulgates bills as a reactionary step in most circumstances. Most, if not all, laws were introduced and passed in response to serious concerns like cyberstalking, cyberbullying, revenge porn, kidnapping, or child pornography.

There is also the possibility of "stealing" or "conversion" in the virtual space." We should not be fooled into believing that we cannot lose money or possessions simply because we are in the so-called digital age. When a victim's finances or assets are taken without their agreement, this is called theft in civil courts. In the Metaverse, there is likely less of a probability of committing theft now. Nevertheless, we are convinced that criminals will find a method to escape.

Requirement for Metaverse in the Future

There is a lot of anticipation for the future of the Metaverse! However, there is still more work to make it a reality for people, companies, brands, and communities. Several essential aspects must be considered to create a vibrant and flourishing Metaverse that provides advantages and opportunities to the greatest number of people.

Next Wave Digitization

Since the 1990s, there has been a considerable surge in digitization. The first 2G cellular network was launched in 1991, and decades after, 5G networks increased digitally. More of our events, settings, and things will need to be swiftly digitized in many and varied media forms due to these improvements.

The digital twin is a concept used to denote both digitized and born-digital content. A rise in the digitalization of historical components is also necessary for the Metaverse to capture creation in our changing present and future culture. This need is driven by the desire to deliver content, goods, and services to clients on-demand and anywhere they can access the internet at a low cost.

Companies that have not undergone a significant digital transformation in all elements of their culture, staff, goods, activities, and services risk becoming irrelevant, obsolescent, or even extinct. The time has come for a company to start investing in introducing its goods and services to digital marketplaces. With radical shift and reprioritization, it is still possible to meet up and engage in the Metaverse.

Interoperability and Portability

Avatars, 3D models, AR, VR, MR, XR, and spatial settings are just a few of the asset types that are fast evolving due to market innovation. Asset classes combine with information to create content bundles that fill the Metaverse's numerous platforms. Most of these technologies are likely to be exclusive or platform-dependent. While these platform-specific and private methodologies may rapidly expand innovation, the Metaverse will necessitate product and content portability and interoperability across platforms.

Published and open documentation, continual traveling unique identifiers, direct asset transfer from one platform to another without extensive third-party services, and shared portability, which allows groups of participants to communicate and interact together, are all important elements of portability.

Interoperability allows assets and data to be shared across platforms and networks. There are various principles, but they do not yet account for all of the new media kinds that exist now in a comprehensive and integrated fashion, much alone equip us with assets and data types that may emerge in the future. Existing standards also do not cater to massive amounts of unstructured data. It also fails to account for digital assets of various sorts using minimally feasible and simple techniques.

There is no shared data framework for the Metaverse space, and there is not even one for current systems. However, if accepted by others in an enlarged transparent landscape, an improved shared data framework, like the one created by Microsoft, might figure out the next step toward a more comprehensive approach to data and metadata. More devoted collaboration and cooperation across the business, government and nonprofit players, like with the Open Data Initiative in the past, must be promoted to produce open standards quickly. The Metaverse must be cross-platform to be available to as many individuals as possible. It must function fluidly across partners and platforms.

Global Commons of the Metaverse

The Metaverse may also require a new global commons based on a globally operating system such as Creative Commons legal tools, both human and computer friendly. A commons, or a collection of shared resources that everyone may add to and take from, is critical in the Metaverse's connection with marketplaces as a driver of ideas. Individual artists, companies, and communities will all have to agree to unified policies and vital financial support for the Metaverse's commons. A commons is a gathering space where we may share, cooperate, and collaborate.

At the same time, the Metaverse's global commons must be a springboard of possibilities that allows for self-expression, identity, and expandable prospects for self-actualization. A viable global commons in the Metaverse will need dialogue, judgment, compassion, and trust to be effective. The Metaverse will be built upon open access. The Creative Commons Attribution license and Creative Commons Zero Public Domain Dedication are the legal instruments better placed to facilitate open Access in the world Metaverse at this moment since they permit economic reuse, remixing, distribution, and new content development.

Crucible Networks, Outlier Ventures, and the Open Metaverse Interoperability Group are organizations and entities working to develop an open Metaverse. Crucible Networks' goal is to establish "open Metaverse architecture" by developing common standards that enable individuals to utilize the same avatar and digital identities across mediums. The Open Metaverse OS was designed by Outlier Ventures. It is a decentralized OS based on Decentralized Finance, non-fungible tokens, and crypto. Finally, the Open Metaverse Interoperability Group explores interoperable identities and designs standards for social networks and inventories to bridge virtual environments.

Standards Required

Whether it is referred to as the spatial web or the Metaverse, the technology will require standards. IEEE and the Spatial Web Foundation have declared partnerships and support for full standards to allow a legally-aligned 21st-century "cyber-physical" web. Since the spatial web is the next phase of global network technology, this is a significant step forward. The spatial web will unleash smart Cities and the Metaverse. An ultra-connected, contextually conscious network of people, artificial intelligence, and computers will enable the creation of digital twins and other forms of digitalization.

This collaboration set of Metaverse standards was produced not simply by developers but also by privacy activists, technology experts, and cybersecurity specialists. Hyperspatial Domains, Hyperspatial Transaction Protocol (HSTP), and Hyperspatial Modeling Language (HSML) are among the spatial web Foundation requirements that regulate context-conscious collaborative technology that guarantees data authenticity data lineage, and data interoperability across integrated hardware.

The XR Safety Initiative (XRSI) has also created a collection of standard definitions for the extended reality space (an umbrella term for all interactive systems, including augmented reality and virtual reality) and a new collection of standards for diversity inclusion ethics, and safety. XRSI recently published a concept for privacy and safety in extended reality and is now working on version 1.1, which will broaden the scope to include the Metaverse. Furthermore, this non-profit is counseling several United States lawmakers and senators on security, privacy, and trust for extremely sensitive sectors like Immersive Healthcare. They are also collaborating with the Australian eSafety Commissioner and the British National Health Service.

10. METAVERSE AND ENTERPRISES

How can the Metaverse Help Enterprises Businesses?

Because of the Metaverse's immersive, fully-accessible, and open character, it is ideal for reproducing crucial human interactions in a virtual environment. The fact that these contacts might extend beyond recreational and social activities should not be overlooked. Those businesses that are starting to look toward a new future for the workplace are also thinking about how the Metaverse may impact their operations.

Many corporate apps for the Metaverse will most likely begin life as upgraded versions of consumer-oriented solutions for their respective markets. Companies, for example, may construct wonderful virtual worlds in which to develop goods, test ideas, and collaborate on innovations with colleagues. Working in virtual reality conference rooms and having interactions improved by MR and AR in the office is becoming more popular.

An increasing number of businesses across various sectors are already using extended reality and Metaverse ideas to bring professionals and workers together in a hybrid environment.

The Metaverse will also open up even more extraordinary chances for online learning and training in the workplace, enabling individuals to immerse themselves in one-of-a-kind experiences that will allow them to develop muscle memory and acquire new abilities as they go. There's also a possibility of establishing virtual landscapes where individuals may more quickly use tools for optimizing their productivity, regardless of whether they're in or out of the workplace.

Microsoft and Accenture have previously experimented with creating virtual workplaces for collaboration, while Facebook (Meta) Horizon offers opportunities to work virtually using headsets and other technologies.

Customer and Business Relationship

Customers can take a tour, engage with a space, or interact with 3D objects such as vehicles, jewelry, or any digital asset by visiting Vstores, or virtual showrooms. Customers will be able to use augmented reality to try on glasses or makeup, as well as picture furniture and other things in their homes, using virtual try-on. Concerts, art festivals, and sporting events are being transformed into digital experiences by creative marketers.

Avatars Require a Budget for Fashion

In virtual reality, selling digital representations of products is becoming a new business stream. As avatars now require a wardrobe change based on events, seasons, and emotions, Gucci, Nike, and Dior all offer digital items that allow personalizing and customizing avatars, whether purses, shoes, hats or sunglasses. Going direct to the avatar (D2A) is a business approach in which marketers sell to digital identities directly.

In a digital environment, designing, making, and selling means there are no shipping or supply chain difficulties. Individuals can use their digital selves to investigate their identities.

It's Game Time for the Brands. It's a Case of Meta-Branding

Sponsoring events in the physical world has shown to have a positive return on investment, and this standard marketing strategy can simply be applied to the Metaverse. Gamifying commerce entails increasing the level of innovation in the competition.

For example, to engage a younger audience, Louis Vuitton designed a video game that was gamified with branded NFT artifacts. Nike employs 3D technology to allow customers to build and design their own items, gaining vital data in the process.

Virtual Pop-Ups Allow for Engagement from Afar

Using virtual worlds with brand placement to generate a creative and personalized opportunity, especially in post-pandemic periods, is a creative and customizable opportunity. Coach, Disney, and Keith Haring collaborated on a fashion, lifestyle, and art project. The three created a virtual pop-up shop where visitors could learn about and purchase limited-edition physical and digital products. Mickey's ears, shearling jackets, totes, and sweatshirts emblazoned with Haring's signature artwork, as well as AR filters and a bespoke Spotify playlist, were available to purchase through the internet shop.

For Fewer Returns and More Direct Sales, Use Try-on in a Snap

Snap allows advertisers to use Snap's AR platform to engage Snapchatters in "try-on" sessions with shoes, sunglasses, and hats.

They can purchase the goods right away if they like them. *Gucci* served as a case study in this regard.

"The shopping AR experience resulted in a favorable return on investment (ROI), as Snapchatters purchased products directly from the app!" According to Snap, "not bad for a campaign that was merely designed to raise awareness and interaction."

New Currencies

According to William Quigley, the co-founder of cryptocurrency Tether, NFTs may create a more sustainable business source for the Metaverse than the existing one.

NFTs will be the primary source of income for the Metaverse, he said.

"Virtual things are now the primary source of income for the video game industry, generating $175 billion in sales yearly. The Metaverse should be several orders of magnitude larger since it encompasses everything, not just video games.

While NFTs may now be considered a novelty, the possibility for this technique is vast. NFTs may tokenize everything from tangible goods to abstract notions like property rights or the right to vote, from gaming to VR to virtual real estate.

Enterprises Employee Collaboration

The latest pandemic has shown that workers are no longer satisfied with being in offices or 9-to-5 workplaces, nor with having to board on endless commutes to work. At the moment, more people are settling for remote jobs, embracing companies that allow people to work from home, and discarding the traditional career path in favor of freelance or entrepreneurial ventures. When you decide to work from anywhere in the world, the Metaverse theoretically becomes the answer to how we assemble and connect for business purposes, especially on a global scale. Ranging from virtual offices to interviews, many things can be done digitally instead of in person. The Metaverse can easily replace those boring team meetings from 2D into 3D, and allow people to place themselves in a virtual office and feel like they're sitting next to their colleagues. If the Metaverse can be incorporated into the current office software and technology, then anyone can seamlessly host a PowerPoint presentation offline and present it to clients and colleagues in a Metaverse boardroom.

With the Metaverse, a person can easily hand a file to a colleague, which can be downloaded at home when needed. With Metaverse, the condition of work will change since workers can seamlessly work from home, get rid of the commuter, and spend more quality time with family and friends. Although it's easy to imagine how comforting this new setting is, are the companies ready to lose their cubicles forever? Regardless of the advantages of remote work, in cases of staff morale, job regulation, and retention, several businesses may still face challenges bordering on productivity, complete accountability, and time management. Remote work and online communication are becoming rapidly normalized.

The future of work may entail a space beyond that of Google Meet and Zoom. Big tech companies have speculated that a Metaverse will give rise to the possibility of new forms of work known as the infinite office and reshape the digital economy. Newer technologies have provided workers with several convenient alternatives to traditional work, and Zoom has transformed the remote and hybrid work setting. Still, the Metaverse development may easily become the ultimate game-changer. The likelihood of putting on a pair of glasses and entering a new and virtual world isn't just an exciting prospect for entertainment and gaming purposes, but the idea of the Metaverse shares a deep connection with virtual reality and gamification. Virtual worlds like Roblox, Fortnite, and Minecraft already have billions of users and economies that rival can small countries.

The Metaverse will easily impact how we live and work and play a significant role in the world as time goes on. People will learn easily, adapt and socialize more on a Metaverse. Studies have shown that soldiers, surgeons, and astronauts have been training for decades using virtual reality. Virtual reality technology has grown into an extensive general use where employees working in industries such as retail, customer services, and logistics are practicing with virtual reality headsets. It has been speculated that in less than 15 years, virtual reality will be adopted in over 23 million jobs globally—this explains humanity's entrance into the Metaverse.

The Metaverse will help people organize the way they work and broaden the scope of society to become more productive. Shared standards and protocols will bring contrasting virtual worlds and augmented realities into a single and open Metaverse that will aid people to work together and lessen duplication of effort.

Tech companies are investing a lot into building the Metaverse, and Mark Zuckerberg predicted that the Metaverse would give rise to an "infinite office" and also reshape the digital economy.

Possible Application of Metaverse

The first concern of the universe is arguably the major technology required since working in the Metaverse requires a combination of AI, full-body gear and sensors, sophisticated headsets, and powerful cloud connectivity, which may take several years to develop. Also, there are various concerns around security, as a persistent Metaverse may have to capture and store users' data to be able to provide an intuitive experience. Furthermore, social issues, such as discrimination and sexual harassment in a Metaverse workplace, must be duly explored in comprehensive detail before large-scale deployment. Currently, working in the Metaverse is possible in a limited way if a person already has the gear and a team on board. For many users, work is their only application of the Metaverse, as a disconnection from the real world may not be worth it without a direct and correlated outcome. Stanford professor and VR expert Jeremy Bailenson stated that he doesn't use VR for recreation since it's not something that a person does for fun yet. Undoubtedly, virtual reality has always been about solving difficult and complex problems. Hence, remote work is a hard problem that VR could help to solve in the world today.

The Metaverse is a repetition of the internet that replaces staring at 2D screens with an immersive digital world that people can access through hardware like VR goggles or even smartphones. For Facebook, a Metaverse includes diving deeper into developing communities, though in a 3D virtual world where the users adopt avatars. Historically, the idea of a Metaverse existed long before Mark Zuckerberg's announcement in 2021. The term was first coined in the 1992 dystopian sci-fi novel, *Snow Crash*, written by Neal Stephenson, and the business applications have already been under development within B.C. Below are some of the top business applications for the Metaverse;

1. **E-commerce:** This will be a huge cost-saving, and will facilitate a reduction in carbon footprint, according to Ashley Crowder, CEO of Vntana, a platform for developing 3D digital and augmented reality e-commerce. He further stated that taking this a step further, the Metaverse will offer a whole new revenue stream for digital goods. Therefore, fashion, furniture, and other companies will not only profit from an increase in sales by using 3D but can also sell the digital assets in the Metaverse.

2. **Legal purposes:** There is a potentially much more pronounced and significant impact that the concept of the Metaverse will generally have on the way business is transacted, according to Jack Newton, CEO of Burnaby legal-technology firm *Themis Solutions Inc.* Though important client meetings are often done in person, wider adoption of the Metaverse will create a potentially zero need for that in the future.

3. **Human resources:** In organizations, such as *Boeing* or *Hewlett Packard*, and anything that includes complex assembly of machinery and would have required a significant amount of training and resources will be

done with virtual reality, augmented reality that are overlaying instructions in a real-world environment and providing a person with a step-by-step playbook on how to put together a jet engine.

4. **Sales/Marketing:** From a sales perspective, experts foresee the Metaverse changing how companies spend marketing dollars. Thus, the cost of advertising may become more distributed among individuals rather than being focused on platforms.

5. **Decentralized Finance:** The interest in cryptocurrencies and non-fungible tokens (NFTs), which are both backed by blockchain to ensure security and authenticity, have rapidly increased throughout the pandemic. Wider use of the Metaverse will see users navigate through virtual worlds, interact with other people and make purchases online that depend on Crypto and NFTs. Also, lending and exchanging such assets in these virtual worlds is predicted to expand.

Risks for Enterprises Businesses Adopting the Metaverse

The thought that certain firms engaged would most likely acquire personal information from users via wearable devices and user interactions is an increasing source of worry among consumers. Another is that Facebook intends to continue to use targeted advertising inside the Metaverse, which has sparked new concerns about the spread of disinformation and the loss of personal privacy.

Another area of worry in the Metaverse is users' addiction

to social media and social media usage in a harmful manner. Internet addiction disorder, social media addiction, and video game addiction are all conditions that may have serious emotional and physical consequences if left untreated for an extended length of time. Depression, anxiety, and even obesity are still a source of concern for some professionals today. They are also concerned that the Metaverse may be exploited as an "escape" from actual reality, isolating individuals from tangible social relationships.

METAVERSE PROJECTS

Project 1

Enjin (MetaCity)

MetaCity is the world's first free-to-earn NFT real estate, influenced by the game *Minecraft*. MetaCity currently has only 70 NFT plots accessible for owners to create a variety of firms.

MetaCity is one of the popular games because of the free NFT drops that you may earn, market, or keep forever. This implies that customers will be able to swap stuff across the two games, MetaCity and Minecraft, in this case.

By simply doing what you enjoy, you add value to the world and make a living. In the MetaCity, for instance, anybody will be able to: Create non-fungible token characters and non-fungible token art exhibitions.

Have you heard of the wildly popular *Grand Theft Auto* server game? Did you know it is now possible to play it in the Enjin Metaverse? MetaCity's Fractured Lands NFT properties became live recently. The benefit of this platform is that it allows users to spend their cryptocurrency assets within MetaCity. This implies that participants can claim a piece of land, and a non-fungible token will be created to show its worth.

Finally, anybody may begin playing the game for free and collect capital coins to purchase NFT plots on the Minecraft server. Currently, just that server is available. It is vital to note that purchasing NFT plots is currently unavailable. This function will be available shortly.

Project 2

Sandbox

The Sandbox is a virtual environment built on the Ethereum blockchain network, where gamers create, control, and monetize their gaming time. The idea is to upset established game producers such as *Minecraft* and *Roblox* by giving creators true ownership in the form of non-fungible tokens and compensating them for their contributions to the environment.

The Sandbox is one of the top 5 Metaverse projects, according to *CoinGecko*, accounting for 7% of the market. The Sandbox's framework for creating games within the Metaverse is one of the reasons you will like it. Users who create content using blockchain networks and smart contracts will have a better experience. The Sandbox Game Maker exemplifies it.

The Sandbox also contains a tool called VoxEdit that allows anybody to begin producing objects. Imagine how much fun it would be to make 3D assets. VoxEdit allows users to post, publish, and sell NFT creations.

Once completed, assets can be sold by posting an initial offering on the NFT marketplace, where interested purchasers can bid on them. In addition, The Sandbox will shortly launch a limited-time experience called "Main Hub." This will be a gathering spot where people may mingle. We are sure you have seen the series "The Walking Dead." On the other hand, this franchise has purchased a property in Sandbox to provide a fantastic experience for its customers. Follow this guide if you wish to buy land in the "Walking Dead" zone.

Snoop Dog is in The Sandbox, which is one piece of evidence that this Metaverse is growing in mainstream popularity. He will have a residence and an exclusive NFT collection to conduct live concerts and interact with gamers.

The Sandbox's governance token, $SAND, serves as the transaction's backbone. This enables participants to, among other things, access the site, play games, stake money, and receive prizes. More than 20 exchanges have listed the $SAND token. Crypto.com, Polionex, Bittrex, and Kraken are just a few of the most well-known exchanges.

How to Participate in the Sandbox?

By the end of the year, the Sandbox will be released. Nevertheless, you may now purchase a parcel on OpenSea in the meantime. The pre-season 0 activities will be available first to landowners. Other players will be able to pre-register and gain entry to the event as time goes on.

Go to The Sandbox's official website and look at the Map section to get started. On the first day of the activities, much of the map will be obscured by fog. The fog will be lifted, and experiences will be unlocked systematically throughout the event for several weeks. The Sandbox Marketplace also has a large selection of non-fungible tokens to choose from.

Project 3

Ultra

Ultra is the pioneer fee-free blockchain network ecosystem of its type, enabling an entertainment platform that brings together diverse gaming sectors and blockchain-driven services in one place. Ultra's network will be able to handle over 12,000 transactions per second.

Ultra will offer you access to a wide range of centralized and decentralized services, including the ability to find, purchase, play, and sell games and in-game stuff, watch live-streaming feeds, communicate with your desired influencers, enter contests, participate in tournaments, and many more.

Ultra, however, has its own ERC-20 coin, $UOS. Uniswap, Bitfinex, Bitrue, Kucoin, and Bancor Network are all places where you can buy $UOS. Transactions on Ultra are also free and very instantaneous. Anybody can stake $UOS and get exceedingly limited edition NFTs as a reward. When the system becomes congested, users' transactions are placed in a queue that automatically prioritizes them. The users who have invested the most $UOS tokens will be moved to the head of the line.

How to Participate in Ultra?

Ultra is still in beta. The official release, however, will happen shortly. The beta version contains 3 stages that must be completed in that sequence. The first phase started in December 2020. The Ultra Platform is a downloadable program that includes an Ultra wallet and access to the Ultra blockchain-powered application ecosystem. Participants will be able to download the wallet-only software after the Mainnet has been connected to the Ultra platform.

All participants' favorite programs may be found in one spot. They interact with entertainment services utilizing future DeFi apps in various ways, from playing games to trading on non-fungible token marketplaces. To open an account today, you must first download the wallet.

Conclusion

Suppose you are still trying to wrap your brain around virtual reality, augmented reality, or mixed or extended reality ideas. In that case, it is time to move on because more tech firms are talking about a new era of the internet known as the Metaverse.

Despite its ambiguous definition, the word "Metaverse" is commonly used to express the idea of a future iteration of the internet consisting of shared, 3D virtual places linked into an imagined virtual world. In basic terms, the Metaverse is a virtual environment where you may use avatars to interact with other individuals in different geographical locations. Tech firms are considering various sorts of Metaverse platforms. Users will buy land and develop ecosystems using NFTs and cryptocurrencies on a blockchain-based platform. A robust virtual environment where individuals may work, play, or interact could be another sort of platform. However, creating a Metaverse will rely heavily on AI and machine learning, as the Metaverse's goal will be to combine our physical world with the virtual space via avatars. The Metaverse will be the convergence of virtual and augmented reality.

The Metaverse development will affect every facet of our civilization, especially entertainment, advertising, and the economy. The Metaverse, however, will have legal ramifications. Collaboration and interoperability between Metaverse developers will be one concern. Intellectual property rights will be another issue, which is the case with most virtual products.

Even if the Metaverse currently falls short of the long-term vision that many have for it, it has the potential to alter how we interact with the virtual environment dramatically. A collaborative virtual experience, similar to NFTs, might open up new options for creators, entrepreneurs, gamers, tech firms, investors, and artists, reorganizing and inventing the digital economy.

Whether it's a 2D web page or a 3D virtual world, the action of a widely recognized, representative, and authoritative organization to impose a reference standard with the participation of the main public and private stakeholders is essential. It is said that the meta-universe will belong to everyone, but at the same time, everyone is rushing to develop to seek a competitive advantage.

It should also be noted that the W3C era has undergone profound changes. Thirty years ago, it was in everyone's interest to create an environment that would allow new businesses to develop; no one would have thought that the Internet would get us where we are today, completely subverting social and economic patterns on a global scale.

The revolutionary momentum of the Internet has gradually given way to the dominance of large technology companies that occupy a strong position and are absolutely interested in creating new business platforms but will not risk losing the hegemony they enjoy. On the other hand, the government lags because technology continually works faster than regulatory design.

Aware of the risks and opportunities that will arise, it is necessary to find a basic condition of balance to make the Metaverse a real development opportunity for all and the real manifestation of the dystopian scenarios described far.

Thank you for making it to the end of this book; we hope it has been informative and able to provide you with all the tools you need to achieve your goals, whatever they may be.

This book has tried to highlight all the important points so that you have a general and, in some cases, even specific knowledge of what will be the future of the Internet and of society as a whole in the 21st century, so that you can benefit in advance from the future technological revolution while limiting its negative effects.

I hope this book will really help you achieve your goals.

BOOKS PUBLISHED BY PHAROS BOOKS

MORE BOOKS PUBLISHED BY PHAROS BOOKS

AVAILABLE ON

amazon.in

Flipkart

Plot No..-55, Main Mother Dairy Road, Pandav Nagar, East Delhi-110092